The School of the Americas

AMERICAN ENCOUNTERS/

GLOBAL INTERACTIONS

A series edited by Gilbert M. Joseph

and Emily S. Rosenberg

✛

This series aims to stimulate critical perspectives and
fresh interpretive frameworks for scholarship on the history
of the imposing global presence of the United States. Its primary
concerns include the deployment and contestation of power, the
construction and deconstruction of cultural and political borders,
and the complex interplay between the global and the local.
American Encounters seeks to strengthen dialogue and
collaboration between historians of U.S. international
relations and area studies specialists.

The series encourages scholarship based on
multiarchival historical research. At the same time, it
supports a recognition of the representational character of all
stories about the past and promotes critical inquiry into issues of
subjectivity and narrative. In the process, American Encounters
strives to understand the context in which meanings related to
nations, cultures, and political economy are continually
produced, challenged, and reshaped.

The

SCHOOL

of the AMERICAS

✺

Military Training

and Political Violence

in the Americas

✺

LESLEY GILL

DUKE UNIVERSITY PRESS

Durham and London

2004

© 2004 DUKE UNIVERSITY PRESS

All rights reserved

Printed in the United States of America

on acid-free paper ∞

Designed by Amy Ruth Buchanan

Typeset in Dante by Keystone Typesetting, Inc.

Library of Congress Cataloging-in-Publication

Data appear on the last printed page

of this book.

CONTENTS

✦

LIST OF ILLUSTRATIONS

ACKNOWLEDGMENTS

✦

Finishing *The School of the Americas* finally gives me the opportunity to express my gratitude to everyone who supported or participated in my
research, and sitting down to thank them is one of the most enjoyable
aspects of the long research and writing process that led to this book. It is
not easy to find the right words. I should begin by saying that an enormous
number of people—more than I can possibly thank here—identified with
the project and aided me because of their own commitment to social
justice in the Americas. Meeting them was a very special experience,
because they drew me out of the self-absorption and isolation that often
afflicts the world of academia and provided inspiration when I most
needed it.

Several colleagues read the entire manuscript or portions of it, and
their comments have greatly improved the final version. Andy Bickford
read and then reread more dyslexic drafts than I care to remember. He
was a superb advisor, and the conversations around my kitchen table have
not been the same since he and Nia Parsons left for Berkeley and Santiago
respectively. Marc Edelman's careful reading of the manuscript and his
wide-ranging knowledge of Latin America strengthened the book considerably. There are few scholars who combine Marc's intellectual brilliance
and dedication to social justice issues, and I am a better anthropologist for
having had him as a friend and critic since our graduate school days at
Columbia University. Hanna Lessinger dissected several chapters with her
red pen and made them more coherent as a result. She also raised a
number of important issues that I would not have addressed without her

advice. Antonio Lauria, Cathy Schneider, Brian Ferguson, Angelique Haugerud, and Patricia Silver offered sound guidance on several chapters and assured me that I was on the right track. Finally, Art Walters told me and anybody who paid attention that I was smart. He also listened to many of the arguments in the book and worried about me when I did not. His love and support have meant a great deal.

This project involved a lot of travel in the United States and Latin America that I could never have done without the generous support of several institutions. A Research and Writing Grant from the John D. and Catherine T. MacArthur Foundation supported me for a year and a half. It was preceded by a grant from the Harry Frank Guggenheim Foundation and a University Senate Research Award from American University. Many thanks to these organizations for their generosity. I also had the good fortune to spend a month at the Rockefeller Foundation's Bellagio Center in Bellagio, Italy, as I immersed myself in writing the book. My warmest thanks to the staff and the other fellows for a wonderful experience. Special thanks to Kathleen Cahill for the Shakespeare quote in the prologue.

Dale and Ken Smith of the Magnolia Hall Bed and Breakfast enhanced my trips to southwest Georgia with their incredible hospitality. They went out of their way to accommodate me, especially on my extended visits, and they helped me to better understand social life in the South. Dale and Ken made me feel like a member of their family, and I will not forget their generosity.

Marc Edelman and Joe Eldridge provided me with contacts and guidance before I left the United States for Honduras, and, in Tegucigalpa, Berta Oliva, Letícia Salomón, and Leo Valladares helped to orient my thinking about the military and human rights. I am also indebted to Ritza Romero of the U.S. Military Group for arranging hotel accommodations, setting up interviews with several SOA graduates, and taking me around town.

George Ann Potter opened her home to me in Bolivia and facilitated my travels in the Chapare. She also taught me a lot about the failings of "alternative development," as it is currently defined. Jaime Zambrano and Juan Carlos Coca Morales accompanied me on my first trip to the Chapare, introduced me to peasant leaders and coca growers, and shared their considerable knowledge of the region with me. I really enjoyed their company. The U.S. Military Group in Bolivia, especially Lieutenant Colonels Rand Rodríguez and Oscar Pacello, expedited my interviews with SOA graduates, and I appreciate their time and effort as well.

I am very grateful to Witness of Peace and Gail Phares for organizing a delegation of U.S. human rights activists to visit the Putumayo region of Colombia in January 2002. It was a privilege for me to join this group. I learned more about a war zone and the life-and-death struggles of some incredibly brave Colombians than I thought possible, and I was also encouraged by the U.S. labor and environmental activists who accompanied me. Their commitment to a more humane U.S. Colombia policy offers an incentive to social scientists seeking ways to connect their research to progressive political change. On an earlier trip to Colombia, Pilar Cubillos of the U.S. Military Group scheduled my interviews with SOA graduates, when she had more pressing things to do. I enjoyed meeting her and value her assistance.

Jeremy Bigwood shared declassified documentation that he acquired through the Freedom of Information Act, and he and Linda Panetta provided their photographs for the book. Laura Catrelli, Megan Fuller, and Beatriz Oropeza did an excellent job transcribing hours of taped interviews, and I enjoyed their commentaries about what they heard. Although the bulk of my fieldwork employed the tools of ethnography, I spent a considerable amount of time combing through various archival collections, seeking information about the School of the Americas when it operated in Panama. I would like to thank Kevin Simons and Carlos Osorio of the National Security Archives, Richard Boyland of the National Archives, and the staff of the Washington National Record Center for their help in locating documents.

I wish to sincerely thank all the anti-SOA activists around the country who shared their experiences and information with me. While she was a student at American University, Kate Lowe introduced me to other young activists, shared her organizing experiences with me, and energized my Latin America class. I am especially grateful to Father Roy Bourgois and Charlie Liteky, who always found time to answer my questions, either in person or over the telephone. Roy's limitless energy and boundless enthusiasm were remarkable, and Charlie's journey from Army chaplain and Congressional Medal of Honor recipient to war resister and anti-SOA activist never ceased to amaze me.

During the year that I spent writing this book, I profited from the opportunity to discuss my ideas with the talented group of young Swedish scholars who organized the Living Beyond Conflict seminar at the University of Uppsala. In particular, I would like to thank Mike Barrett,

Chris Coulter, Staffan Löfving, Sverker Finnström, and Charlotta Wid-mark for their hospitality and for sharing insights from their own important research with me. Thanks also to Björn Beckman of the International Development Studies seminar and Mona Rosendahl of the Latin American Institute at the University of Stockholm for organizing a very lively discussion of my work at their university.

This book would not have been possible without the participation of the students and staff of the School of the Americas and members of the Colombian, Bolivian, and Honduran armed forces. Thanks to all for talking with me. Colonel Glen Weidner, who is now retired from the U.S. Army, opened the School of the Americas to me, and he initiated my contact with the militaries of Colombia, Honduras, and Bolivia. I value his assistance, and I also appreciate the many hours that he spent answering my questions, when he did not know what I would write. Glen Weidner and the Bolivian major I call Juan Ricardo Pantoja dissolved military stereotypes for me, and they helped me to grasp some of the complex human qualities that characterize the men who serve in the armed forces of the Americas.

Finally, this book is dedicated to the memory of those Latin Americans who were murdered, tortured, and disappeared by the security forces, and to the peasant coca growers of the Andes who fight against incredible odds to live in peace and dignity. My hope is that the book will contribute to building the kind of world that so many of them have struggled to create—a world based on equality, justice, and accountability.

❖

The Teflon Assassin

Meet it as I set it down
That one may smile and smile and be a villain.
—William Shakespeare, *Hamlet*

Luis Bernardo Urbina Sánchez. So this is him, I thought to myself, not knowing what to expect. I had been anticipating our meeting for several days—wondering what he was like, how I would interview him, and why he would agree to talk to me about his career. Across from me sat the retired Colombian general, an alleged murderer and paramilitary coordinator during his long cold war career in the army.[1]

The fifty-eight-year-old man did not look like an army officer, much less someone accused by international human rights organizations of terrible crimes. He wore dark slacks, a sport jacket, and a tie. Longish black hair speckled with gray curled from beneath a cloth cap and covered the nape of his neck. A neatly trimmed, full beard softened the lines of his face. Tortoiseshell glasses framed his large brown eyes, and a pleasant smile spread across his face when he shook my hand. There was nothing unsettling about this attractive, middle-aged man. How, though, does one recognize a killer—dark aviator glasses, a pencil-thin mustache, a permanent scowl, an arrogant swagger, and a protruding gut? Urbina possessed none of the stereotypical features of the brutish Latin American army officer. He reminded me less of the stiff, uniformed men who strode the corridors of the Colombian defense ministry than of my male colleagues in academia.

He glanced around the room as I fumbled with my tape recorder, but he made no attempt at small talk. Folding his hands on the table, he sat up straight, smiled, and waited for me to begin. Urbina was the last of a series of Colombian, Honduran, and Bolivian alumni from the U.S. Army's School of the Americas (SOA) whom I interviewed during the summer of 2001. He was doing his old friend, Nestor Ramírez, a favor. The two men had trained at the SOA together in 1985, and General Ramírez—now the second-in-command of the Colombian armed forces—had asked him to talk to me. Ramírez wanted to showcase Colombian SOA graduates who had stellar careers in the top ranks of the military, but the human rights violations attributed to Urbina were apparently irrelevant to the commander and left him unimpressed; indeed, Ramírez himself stood accused of a 1986 revenge slaying that was never investigated.[2] Perhaps Urbina, for his part, also wanted to use me to proclaim his innocence. He would soon tell me that his hands were clean and that he had nothing to hide.

"So why don't you tell me about your military career," I asked.

Luis Urbina did not want to be a soldier, he said. His father owned a modest cattle ranch and had worked hard as a veterinarian in the small Colombian town of Nemocón to support a large family of ten children— five girls and five boys. Luis and his brothers were like other provincial young men of their generation whose families expected them to pursue careers in the army or the priesthood. Luis chose the latter and studied in a seminary until he was eighteen. Then, as he tells the story, he "fell in love with a girl." The unplanned consequences of this romance got him kicked out of the seminary and thinking about a career in the army. It was not long before he was packing his bags and heading to the military academy in Bogotá. His army training lasted several years, and when he finished in the mid-1970s, he began a series of postings around the country.

Urbina walked me step-by-step through his career and his rise through the ranks. He spoke in a matter-of-fact manner that demonstrated little eagerness to enhance his image as a key player in the Colombian army's long war against leftist insurgents. He was engaging and urbane, and he spoke in an easy, relaxed manner. There was no reason to posture: he was, after all, a general, and he knew that I understood the importance of rank.

Urbina rose to head the Department of Administrative Security (DAS) from within the Colombian army's secretive intelligence apparatus.[3] He spent much of his career with specialized units attached to brigades and battalions located in areas where guerrillas of the FARC (Fuerzas Armadas

Revolucionarias de Colombia), the ELN (Ejército de Liberación Nacional), and other organizations were active. The intelligence units and the operatives who commanded them were the linchpins in the military's counterinsurgency campaign, the interface between state security forces and their vicious paramilitary allies—the so-called self-defense forces. Intelligence agents cultivated relations with paramilitary groups and coordinated their activities. Sometimes they worked temporarily with the paramilitaries, but at other times they paid them to murder selected individuals. Information about the guerrillas and their unarmed sympathizers flowed freely between the army and the paramilitaries, many of whom were themselves former military officers.

The human rights violations attributed to Urbina took place between 1977, when he was a captain in the Second Brigade's intelligence unit, and 1989, when he was a colonel in the DAS. They began with the disappearance of Omaira Montoya Henao and Mauricio Trujillo in the coastal city of Barranquilla. The pair was snatched off a street and then brutally tortured; Omaira Montoya was never seen again. Evidence linked Urbina to the crime, but he was not investigated.

In 1986, Lieutenant Colonel Urbina, recently returned from the School of the Americas, managed a regional intelligence network from the headquarters of the Fifth Brigade, located in the highly conflict-ridden Middle Magdalena region of central Colombia. There was nothing high-tech about his operation. "I'm an old soldier," he explained, "so for me you can never replace human intelligence [with technology]." But putting an intelligence network together was not easy. "You recruit people. It is more a question of common sense than training. To teach a person to become a good director of a network is very difficult. You invent things and you get results . . . it takes a long time to really learn how to do the work." During the two years that Urbina spent with this unit, William Camacho Barajas and Orlando García González were detained by an army patrol in the town of San Gil. They were registered under false names, taken to the Fifth Brigade's intelligence unit, and never seen again. Ten months later, Mario Alexander Plazas disappeared; his burned cadaver was discovered in the town of Piedecuesta. A Fifth Brigade intelligence agent confessed to the murder, stating that Urbina ordered the disappearance, torture, and execution of the young man. Urbina, however, was neither questioned nor linked officially to the crime. Shortly thereafter, paramilitaries and army intelligence agents dressed in civilian clothing murdered the mayor of Sabana de Torres, Alvaro Garcés Parra. Urbina was again fingered by an insider as the intellectual author of the crime, but no official questions were asked. The army rewarded Urbina for his service in the Fifth Brigade by promoting him to full colonel.

By 1989, Urbina had earned a ticket to Bogotá and a job in the DAS, where he allegedly coordinated a nationwide paramilitary network that disappeared and murdered individuals identified as guerrillas or guerrilla sympathizers. During his tenure as an army spy chief, he maintained close ties with a former U.S. colleague from the School of the Americas who worked in the embassy. "The Americans," he said, "gave me cars. They equipped units, and they set up communications systems for me. We achieved good results." Indeed, the first Bush administration was assisting the army's counterinsurgency campaign with rising levels of military and intelligence aid. Although this support was earmarked officially for an escalating war on drugs, U.S. officials knew that the Colombian military would cooperate in the drug war only if the aid allowed them to pursue the guerrillas, their main adversary.[4] So when Urbina targeted the insurgents and their civilian sympathizers, the protests of these U.S. policymakers were muted, if audible at all. "The army is the primary entity charged with fighting subversion," explained Urbina, "and we began to hit the subversives using the U.S. aid to fight the drug traffic." During this period, Amparo Tordecilla disappeared, kidnapped by a group of men who approached her in a taxi. Her cadaver and those of others were later discovered in a clandestine cemetery on the outskirts of the city. One of the individuals who participated in the abduction claimed that the taxi belonged to the army and that it was under Urbina's control.

None of this affected Urbina's career in a negative way. In 1991, the government sent him to Venezuela as its military attaché. It had opened negotiations with the FARC in Caracas, and the subject of amnesty for the insurgents was on the table. Urbina, like other military hard-liners, opposed any kind of amnesty that "sold out the country." He manipulated the government negotiators with intelligence that he fed them about the guerrillas and eventually took credit for the collapse of the talks. "They broke off the negotiations and that is what I wanted," he explained. "It was a good action." When he returned to Colombia, the army wasted little time in promoting him to brigadier general. Of the sixty-two men who graduated with him from the military academy, Urbina was one of only five to make general.

Urbina mentioned the Tordecilla case during our interview, making vague references to the matter but then brushing it aside with comments like "I'll tell you about it later." He felt more comfortable blaming human rights organizations for the troubles in which he and other military officials found themselves. According to the general, human rights groups worked in league with the guerrillas, drug traffickers, and common delinquents to attack the integrity of the armed forces. "If a commandant does the right thing in his operations and hits the subversives hard, it's easy to denounce him and to say that he was the one who disappeared such-and-such a person." He then mused about the fate of former Chilean dictator Augusto Pinochet. "Just take the Pinochet case," he said. "Pinochet is paying the consequences of something that he probably wasn't even aware of. The consequences are terrible. Here in Colombia we are heading in the same direction . . . I go to get my ticket to travel to the United States and they have me figured as a delinquent . . . a bandit."

Later in our conversation he referred obliquely to a book published by a group of human rights organizations that delineated a series of allegations against members of the Colombian armed forces.[5] "Because of this book, a lot of people lost their [U.S.] visas," he complained.

"Do you appear in the book?" I asked.

"I don't know," he muttered, looking at the table. "But what I am telling you is that they [guerrillas and human rights organizations] are repersecuting the people who are hitting them. Yes, I am in the book, but I haven't even been investigated for what they say because [the charges] are lies. And they involve my brothers who had nothing to do with what happened."

I pressed him to explain his version of the events in the Tordecilla case, but he offered few specifics. "When I was the director of army intelligence," he began, "they say that a woman who was the lover of a guerrilla commander was disappeared. . . . There were many details. People were saying many things, that there had been an army car. . . . They said that the car that they had seen was one of mine, that the car belonged to the army and was under my orders. I can't always know where all of the army's intelligence cars are. . . . They investigated me but I got out from under the problem in 1999. Now I don't have anything to do with it."

Nowadays nobody sees much of Urbina. He retired from the armed forces in 1995 and, like many of his army colleagues, joined the burgeoning private security business. He sold his services for five years to a private,

U.S.-owned transportation company that operated in Colombia, protecting outgoing airplane cargoes from infiltration by drugs. He was well paid for his efforts. "I bought a good apartment," he told me. He also purchased a new car and was secure enough financially to leave this employment and go into business for himself. With four other retired army officers, Urbina founded his own security firm that specialized in protecting cargoes trucked overland by private entrepreneurs.

He resides in an upscale part of northern Bogotá, an area that has been relatively untouched by the rising violence in Colombia. His wife is worried about his safety and would like to leave Colombia, but Urbina maintains that a retired officer belongs in his country, although he acknowledges that many people would like to kill him. The general maintains a low profile: he does not talk to the press, he avoids public events, the beard and glasses are part of a new look cultivated since he left the armed forces, and when he travels to other parts of Colombia, he uses a false name. Urbina enjoys golf and plays a couple of times a week at a club frequented by other army officers. He also visits the United States from time to time. How, I ask, does he enter the United States without a visa. "I've had a visa for a long time," he says, "and besides, they have certainly realized that [none of the charges] are true. I don't know. I go to the United States with no problem."

Urbina's career embodies the issues of military training, U.S. complicity, human rights, and impunity that shaped the bloody history of the cold war in Colombia and much of Latin America. These issues and their complicated legacy are at the core of this book. Urbina expresses no regrets about the past, despite the numerous, serious allegations made against him by human rights organizations that have presented what they consider to be evidence of his culpability. Impunity is widespread in Colombia, as it is elsewhere in the Americas, and none of the charges against Urbina have stuck. The general is unrepentant. Knocking on the table, he exclaims, "I have done nothing wrong, nothing more than serve my country."

✦

The Military, Political Violence,

and Impunity

I don't see why we need to stand by and watch a country go communist because of the irrespon-
sibility of its people. — Secretary of State Henry Kissinger commenting on the election of
Salvador Allende as president of Chile in 1970.

Chile's coup d'etat was close to perfect. — Lieutenant Colonel Patrick J. Ryan, U.S. Military
Group Commander, Santiago, Chile, October 1, 1973

Today our nation saw evil, the very worst of human nature . . . I've directed the full resources of
our intelligence and law enforcement communities to find those responsible and bring them to
justice. We will make no distinction between the terrorists who committed these acts and those
who harbor them. — President George W. Bush, September 11, 2001

Almost three decades before the attacks on the World Trade Center and
the Pentagon killed over three thousand people, another act of unspeakable horror took place in the South American country of Chile on September 11, 1973. A savage military coup d'etat backed by the United States
overthrew democratically elected president Salvador Allende and unleashed a wave of state-sponsored terror that left thousands of Chileans
dead. The events in the United States and Chile, so different in many
ways, shared two important features: the deaths of thousands of civilians
and the involvement of the United States in training the terrorists. Osama
bin Laden joined the mujahideen guerrillas who were organized, trained,
and equipped by the United States to topple a pro-Soviet regime that
controlled Afghanistan in the 1980s, even though U.S. strategists recog-

nized that many of the Islamic fundamentalists of the mujahideen opposed democracy, women's rights, and religious pluralism.

General Pinochet and his compatriots in the Chilean armed forces were also aided and abetted by the United States despite their use of terror at home and abroad. Almost all of the Chilean officers who overthrew Allende had trained at a U.S. military service school prior to the coup; most had attended the U.S. Army's prestigious School of the Americas, a training institution where Latin American soldiers learn counterinsurgency warfare. The most notorious acts of international terrorism committed by the Pinochet regime included the 1974 car bomb assassination of General Carlos Pratts and his wife in Buenos Aires; the 1974 attempted murder of Bernardo Leighton, the founder of the Chilean Christian Democratic Party, in Rome; and the 1976 car bomb execution of Orlando Letelier, Allende's former ambassador to the United States, and his U.S. aide, Ronnie Moffat, in Washington, D.C. The assassinations were orchestrated by the Chilean secret police and connected to Operation Condor, a network of South American intelligence agencies that collaborated in hunting down and assassinating political dissidents who opposed the dictatorships in their respective countries. The fact that the Letelier murder was carried out in the heart of Washington, D.C., testifies to the confidence with which Pinochet's secret police operated in the United States and suggests that the CIA was probably aware of its activities.[1]

The dual tragedies of September 11 force us to recognize that the United States government has assisted in the creation of international terrorist networks and has rarely let a commitment to democracy stand in the way of its global ambitions. But until the attack of September 11, 2001, American citizens seldom experienced the horror, the anguish, the profound loss, and the lingering sense of vulnerability that the survivors of terrorism in other parts of the world know too well. From Chile to East Timor, Congo, Guatemala, El Salvador, Colombia, and many other cold war battlegrounds, ordinary people who desired land reform, better wages, improved health care, education, and the basic right of self-determination were labeled communists by U.S.-backed regimes and murdered, tortured, and disappeared by shadowy paramilitary death squads and state security forces trained by the United States. The perpetrators were almost never held accountable, and officials acknowledged the dead and the abused very slowly, if at all. The Third World victims of cold war atrocities usually did not receive public commemorations, such as those so fittingly published by the *New York Times* for each individual killed in the World Trade Center, nor were memorials constructed in their honor.

Forgetting the proxy wars and covert operations carried out by the United States and the Third World security forces that do its bidding obscures the extent to which modern America emerged as the result of an imperial project that brutalized and oppressed peoples around the world. To understand these international adventures, a broad conceptualization of imperialism is useful, one that begins with the intrusion of U.S. economic interests into other countries and extends to the multiple and varied practices of political, military, and cultural domination.[2] The empire which the United States now possesses is notable for the constellation of military bases that dot the globe; the defense budget that, even before September 11, 2001, totaled billions of dollars; the stockpile of nuclear weapons capable of destroying humankind; the ongoing alliances with repressive regimes that range from the Saudi royal family to the unrepentant military of Guatemala that rules behind a facade of civilian government,[3] and the history of military intervention that continues unabated, as the invasion and occupation of Iraq so amply demonstrates.

Military bases, weapons, and strategic alliances with local security forces constitute the cutting edge of the U.S. empire in which the American state rules less through the control of territory than through the penetration and manipulation of subordinate states that retain considerable political independence.[4] As Panitch and Gindin (2003, 30) note, "the need to try to refashion all of the states of the world so they become at least minimally adequate for the administration of global [capitalist] order . . . is now the central problem of the American state." This is an enormously complex and difficult task that requires dense networks of economic, cultural, social, and military control; indeed, the frequent inability of the American state to turn subordinate states into effective instruments of U.S.-led global capitalism generates policies aimed at removing the threats posed by so-called rogue states.[5]

U.S. imperialism, however, extends well beyond military interventions, foreign policy debates, and the intrusive economic policies of international financial institutions, such as the World Bank and the International Monetary Fund.[6] The historian William Appleman Williams describes the U.S. empire as "a way of life" (1980). Understood in this way, imperialism unfolds at the local level through a variety of power-laden relationships between unequal social actors. The security forces—militaries, paramilitaries, militarized police forces—constitute one of the most basic forms of imperial intrusion and control, as they enforce the systems of order required by dominant groups to manage different kinds of people. The relationship of Third World security forces to the United States, to each other, and to various noncombatant civilians takes different forms under changing historical conditions.

Dealing with the dark, seamy side of U.S. involvement in global affairs has never been easy for the citizens of the United States because of widespread amnesia about twentieth-century U.S. empire building. A broad cross-section of Americans like to think of their country as a land of freedom, a beacon to the oppressed, an exemplary democracy, and most recently, a righteous crusader against global terrorism. This nationalist vision has deep roots in notions of American exceptionalism and distinctiveness, but U.S. citizens and policymakers cling to it at their own peril. Summarizing an ongoing debate between U.S. diplomatic historians and American studies scholars, an editor of a recent collection of historical essays suggests that

> to argue in the manner of George Kennan and subsequent genera-
> tions of "realists" (and latter-day "post-revisionists") that if the United
> States *briefly* had an empire in the aftermath of the Spanish-American
> War, it promptly gave it away; that, therefore, imperialism has always
> been inconsequential to U.S. history; that, unlike the great powers of
> Europe, the historical experience of the United States has been char-
> acterized by "discovery" not "imperium," "global power" not "impe-
> rialism," "unipolarity" not "hegemony" is to perpetuate false notions
> of "American exceptionalism" and to engage psychologically in denial
> and projection. Such arguments also ignore structures, practices and
> discourses of domination and possession that run throughout U.S.
> history (Joseph 1998, 5–6).

The mystification of U.S. involvement in global affairs is reflected, at least in part, in the naive headlines that asked "Why Do They Hate Us?" on the front pages of U.S. newspapers and magazines in the aftermath of September 11. This ingenuous question suggests that past U.S. aggression never existed, or if it did, it was unintentional. The question also points to the inability of many Americans to move beyond the hopelessly provincial understandings that inform their views of the world's peoples.

Since the nineteenth century, however, the United States has willfully embarked on a career as an imperial power, and it has assembled the tools of repression that this required. Many U.S. citizens have cheered its progress along the way or lived behind a wall of self-absorbed denial and ignorance about the consequences of U.S. foreign policy. As Catherine Lutz has so eloquently written, "we have not evaluated the costs of being a country ever ready for battle. The international costs are even more invisible as Americans have looked away from the face of empire and been taught to think of war with a distancing focus on its ostensible purpose— 'freedom assured' or 'aggressors deterred'—rather than the melted, exploded, raped and lacerated bodies and destroyed social worlds at its center" (2001, 2). This is *not* to claim therefore that the United States deserved what happened on September 11, 2001, or that the perpetrators should remain unaccountable. Such a conclusion mistakes explanation for justification and is itself a product of the historical amnesia at the heart of American nationalism. Nevertheless, in the aftermath of the terror attacks, those who sought explanations in the history of U.S. global involvement were arrogantly dismissed for allegedly suggesting that the country somehow got what it deserved. "Nobody deserves terror," writes Argentine-born Ariel Dorfman, but "what we deserve, all of us, is some measure of justice" (2002, 22).

Justice, however, requires that we distinguish between the civilians who died on September 11 in the World Trade Towers, the Pentagon, and the hijacked airliners, and the high-ranking military officials of the Pentagon who have organized and supported acts of terrorism against innocent people elsewhere. We must also be mindful of the differences between the civilian domains represented by the World Trade Towers and the hijacked passenger airliners on the one hand and the Pentagon, which represents the center of the United States global military apparatus, on the other hand. The perpetrators of the September 11, 2001, attacks must certainly be apprehended and held accountable, but the perpetrators of terrorism within the U.S. military establishment, along with the political leaders who approved their actions over the last half of the twentieth century, are also responsible for their actions and should account for them, if we are to take a consistent stand against organized violence and the deaths of unarmed men, women, and children everywhere.

Investigating and understanding the military's relationships to the peoples around the world long treated as inferior allows us to appreciate how U.S. interventions repressed, terrorized, and humiliated others. It is to comprehend that our grief and sorrow are not unique and that American dead are not the only ones who count. To grasp the complexity of these relationships, we must explore the imperial alliances, social entanglements, networks of power, cultural understandings, and pervasive impunity that have upheld U.S. global hegemony.

This book examines how the United States constructs a repressive military apparatus, in a region long considered by many to be its "backyard," through the lens of the U.S. Army's School of the Americas (SOA). The School of the Americas is a U.S. Army center for Latin American militaries that, since its establishment in the Panama Canal Zone in 1946, has trained over sixty thousand soldiers in combat-related skills and counterinsurgency doctrine. It has been at the center of an intense public controversy over the last decade, because of the participation of some of its alumni in human rights atrocities. Some of the most notorious graduates include Argentine General Roberto Viola, who was convicted of murder, kidnapping, and torture during Argentina's "dirty war" (1976–1983); former Panamanian strongman Manuel Noriega; Salvadoran Colonel Domingo Monterrosa, who commanded the brutal Atlacatl Battalion that massacred nearly one thousand civilians in El Mozote; Guatemalan Colonel Julio Alpírez, who tortured and murdered guerrillas and a U.S. citizen while on the CIA's payroll; and Honduran General Luis Alonso Discua, who commanded an army death squad known as Battalion 3-16.

Critics of the School assert that Latin American soldiers learn the repressive tactics of counterinsurgency warfare, which pits soldiers less against guerrilla insurgents and drug traffickers than against poor peasants and civilians (Nelson-Pallmeyer 2001), but U.S. Army officials identify alumni charged with human rights violations as "a few bad apples" who do not reflect the School's overall success in building ties to Latin American militaries. Because of the controversy, the SOA has gained considerable notoriety. It has been forced to open its doors to greater public scrutiny, and the Defense Department changed its named to the Western Hemisphere Institute of Security Cooperation in 2001.

Starting from the School of the Americas, the book traces the relationships of empire building through the experiences of three groups of people. First, it considers how military personnel from the United States and Latin America engage each other at the School through the quotidian experiences of military training and daily life, and how these highly unequal encounters mold various kinds of relationships, understandings, opportunities, and patterns of collusion that extend across the Americas and anchor a vision of empire in actual experience. It demonstrates how the United States bought the collusion of Latin American security forces in the aftermath of World War II and how, over the course of the twentieth century, it transformed these entities into extensions of its own power in Latin America and internationalized state-sponsored violence in the Americas.

The internationalization of the repressive power of Latin American states reverberated throughout the Americas. State agents became more efficient in carrying out acts of violence, which exacted a heavy toll on the human rights of many Latin Americans. At the same time, the control and influence that the U.S. military exercised over national security forces enhanced the ability of the United States to manipulate independent governments as it pursued key political, economic, and security interests in the hemisphere.[7] Finally, the training and arming of a castelike group of professional soldiers aggravated processes of social and economic differentiation in many Latin American countries.

From the School of the Americas, the focus then shifts to the coca-producing regions of Colombia and Bolivia, where the expansion of the illegal cocaine traffic, the presence of armed guerrillas, and the organization of militant peasant coca-grower unions have led to an intensification of state-sponsored violence. The discussion teases out some of the connections between local-level security forces, the School of the Americas, and other U.S. military training initiatives. It also scrutinizes the consequences of militarization for peasant families and the ways that impunity for members of the security forces and civilian officials shapes the relationship between state-sponsored violence and deepening social fragmentation. The analysis then returns to the United States, where a vibrant social movement dedicated to closing the SOA has focused attention on the School's training practices and connected them to human rights violations committed by School alumni. The discussion examines the challenges posed by the movement and explores the shifting logics of power within the U.S. military as the Defense Department struggles to reconstitute the School, refine its public relations message, and revamp its mission—all in an effort to shore up the legitimacy of a disgraced military training institution and by extension, past and present U.S. policies in Latin America.

How, this book asks, does the United States train Latin American "professional" soldiers who define their agendas in distinctive ways and on whom the United States depends for cooperation but does not entirely trust? How does this training, immersion in the "American way of life," and access to a transnational world of power and privilege shape the Latin Americans' ties to U.S. military personnel, their social and career mobility at home, and their geopolitical understandings? What lessons do soldiers—U.S. and Latin American—draw about the dirty wars that raged across Latin America for the last half of the twentieth century, and to what extent are they now willing to accept, excuse, or condemn the exercise of violence and the violation of human rights? What is the relationship between the "order" produced by security forces in Latin America and the disorder wrought on peasant families, and how does this shape the demand for military training? Finally, how has the U.S. government dealt with its own citizens who oppose the use of their tax dollars for military training and demand that the SOA be shut down? Addressing these questions allows us to move beyond simplistic distinctions between "us" and "them" and to explore the tensions and contradictions that have emerged with the expansion of U.S. military power in the Americas.

Even though the SOA has captured the public spotlight, it is only a small part of a vast network of U.S.-sponsored training programs worldwide. The School graduates between six hundred and eight hundred police and military officers annually, and it trains several hundred more via mobile training teams dispatched to Latin America; in contrast, the United States military instructs some one hundred thousand allied, foreign soldiers in the United States and abroad every year. Approximately forty-eight thousand soldiers and law enforcement officials from around the world trained in the United States in 2000, and between 1998 and 2000, ten thousand to fifteen thousand Latin Americans received instruction from U.S. military personnel in the United States and in their home countries (Lumpee 2002).

In the United States, foreign soldiers and U.S. troops train together in at least one hundred fifty disclosed training centers and military schools, where instruction is geared primarily to the needs of U.S. forces.[8] The international students represent all branches of the armed forces, but most come from the armies of their respective countries. Schools with large numbers of foreign trainees include the U.S. Army Intelligence Center at Fort Huachuca, Arizona, the U.S. Army Command and General Staff College at Fort Levenworth, Kansas, and the U.S. Army JFK Special

Warfare Center and School at Fort Bragg, North Carolina (Lumpee 2002). Several Spanish-language schools, however, operate specifically for Latin American officers. In addition to the School of the Americas, they include the Inter-American Defense College in Washington, D.C., the Inter-American Air Force Academy at Lackland Air Force Base, Texas, and the Navy Small-Craft Instruction and Technical Training School, which operates in Mississippi and North Carolina. Together, they reflect the strategic significance of Latin America for the United States.

The U.S. military and law enforcement agencies train foreign security forces abroad in a myriad of venues. Small Special Forces Mobil Training Teams (MTTS) teach specialized units from numerous countries. They instruct foreign militaries on the techniques for waging small-scale strikes, counterterrorism activities, psychological operations, foreign internal defense (i.e., organizing, training, and advising military and paramilitary forces), "unconventional" warfare (i.e., support of military and paramilitary operations against a standing government), and "such other activities as may be specified by the President or the Department of Defense" (LAWG 1999); moreover, joint training exercises that involve both the Special Forces and the U.S. military's regular forces with their foreign counterparts also take place frequently. In addition, intelligence agencies instruct an undisclosed number of military and paramilitary troops around the world, and the Federal Bureau of Investigation, the Customs Service, and the Drug Enforcement Agency have their own training programs for overseas security. Very little is known publicly about these activities. Finally, a number of private companies, such as the Washington, D.C.–based Dyncorp and Military Professionals International, contract with the U.S. government to carry out military training activities and to maintain high-tech weaponry for Third World clients. They are also directly hired by foreign governments, but their practices are subject to little congressional oversight or public accountability (Amnesty International 2002, iv).

Military training is fueled by an enormous arms industry that requires the availability of training for the continued development, use, and maintenance of weapons. Forty-six of the one hundred largest arms producers in the world are United States companies, and together they sold ninety-six billion dollars worth of weaponry in 2000 (SIPRI 2000). Training is also shaped by the shifting geopolitical field of force in which the United States defines its national interests and security concerns. During the cold war, it was geared to the defeat of the "communist" enemy in the Third World through counterinsurgency programs that combined economic assistance with psychological operations and security measures. Counterinsurgency doctrine was initially geared to defeating revolutionary movements that challenged U.S. hegemony, but during the Reagan administration, a new, more aggressive strategy of intervention called "low-intensity conflict" (LIC) emerged that went beyond targeting insurgencies. The new strategy was to undermine governments that already existed and were perceived to be hostile to the United States. Both counterinsurgency and LIC doctrines advocated similar kinds of economic, psychological, and paramilitary coercion and aimed to defeat any threat to U.S. interests (Klare and Kornbluh 1988).

Fighting "communists"—an enormously elastic category that could accommodate almost any critic of the status quo—became obsolete in the post–cold war era after the collapse of the Soviet Union and the expansion of the drug war in the 1990s. The U.S. military's Southern Command (SOUTHCOM) welcomed the drug war because it enabled SOUTHCOM to expand relations with allied militaries throughout the hemisphere. Counternarcotics training provided the opportunity to strengthen ties to local security forces, and low-intensity warfare strategies employed in Central America were easily adapted to fighting a war on drugs (Youngers 2003). Training programs shifted to so-called operations other than war, such as counternarcotics and counterterrorism activities, although the basic techniques of warfare at the local level remained substantially the same. In the Andean region, the term "narcoguerrilla," which replaced "communist" for a few years, was overtaken by the more ominous-sounding "terrorist." The drug war and the subsequent "war on terrorism" offered convenient rationales for SOUTHCOM to maintain troop and funding levels as other areas of the world, especially the Middle East, became more important to the Pentagon. Following the attack on the World Trade Center, President Bush offered military training to any nation willing to join the United States in a global crusade against terrorism, and the administration shifted the definition of the conflict in Colombia from a drug war to a war on terror in order to justify its involvement in counterinsurgency operations.

A focus on the School of the Americas opens a small window onto the ways that the United States trains diverse foreign soldiers and secures their cooperation. It allows us to explore the creation of coercive, highly unequal relationships between members of the armies of the Americas and to consider how, under the tutelage of the United States, beliefs about professionalism, human rights, just wars, and subversion are crafted. This is important because, according to numerous truth commission reports from the 1980s and 1990s, state security forces were responsible for the vast majority of massacres, murders, disappearances, and extrajudicial executions that characterized the twentieth-century Latin American "dirty wars," when many countries suffered under the boot of military dictatorships (REMHI 1999; ANCD 1986; CNPDH 1994; Comisión de la Verdad 1993), and that continue to plague Andean countries like Colombia. Although limited democracy has replaced military rule, abusive, U.S.-trained armies and counternarcotics police forces are still responsible for most of the human rights violations in Colombia, Peru, and Bolivia, where the United States is involved in a conflict that is at times a drug war, at times a counterinsurgency war, and at times a war on "terror." Militaries retain an enormous amount of political and economic power, and civilian governments have only rarely held military perpetrators accountable for human rights crimes, past and present. When they do, it is usually only after years of struggle by human rights organizations and the relatives of the victims.

In Argentina, for instance, members of the security forces who murdered, tortured, and disappeared thousands of people in a dirty war (1976–1983) benefited from the Obediencia Debida and Punto Final Laws that allowed low-ranking personnel to claim that they were "following orders" and set limitations on the duration of human rights trials. President Carlos Menem (1989–99) then instituted sweeping pardons that exonerated high-ranking commanders convicted of orchestrating the dirty war (Feitlowitz 1998; Verbitsky 1996). Some of these commanders were later retried and convicted for kidnapping the babies born in captivity to female prisoners who were subsequently executed. In Guatemala, most officers escaped prosecution, and by the mid-1990s none had been convicted for ordering the murders and massacres that left two hundred thousand Guatemalans dead during a thirty-five-year civil war (e.g., Schirmer 1998; Perera 1993; Carmack 1988; Menchú 1984, and Levenson-Estrada 1994). In 2002, however, a Guatemalan court convicted Colonel Juan Valencia Osorio for ordering the murder of anthropologist Myrna Mack in 1990. It sentenced him to thirty years in prison, but the glimmer of hope that this ruling offered to human rights defenders was extinguished when the conviction was overturned on appeal. Mack was the founder of a research institute called AVANCSO, which published a report in 1990 that linked the internal displacement of thousands of Guatemalans to the army's counterinsurgency campaign. In Colombia, midlevel officers who tolerated, planned, and took part in paramilitary violence in the 1980s received promotions and currently hold the highest positions in the armed forces (HRW 1996), which continues to wage a brutal civil war in which thousands of innocent civilians are killed and displaced from their homes every year.

In El Salvador, the high-ranking perpetrators of large-scale massacres, such as the one that took place at El Mozote, were not held accountable, and when investigations of human rights abuses took place, the government and the U.S. embassy restricted them to low-ranking soldiers (Binford 1996).[9] The stirrings of justice for some victims only began in 2002, when a U.S. court in Florida ordered two retired Salvadoran generals— José Guillermo García and Carlos Eugenio Vides Casanova—living in the United States to pay fifty-five million dollars to three Salvadoran citizens tortured in El Salvador some twenty years ago, when the Reagan-Bush administration supported the Salvadoran armed forces. Their involvement with El Salvador's dirty war, however, runs much deeper. While he was minister of defense, for example, SOA graduate Garcia failed to investigate the 1980 deaths of four U.S. churchwomen and the 1981 El Mozote massacre. Vides Casanova, who headed the national guard at the time, allegedly ordered the murder of the nuns. Neither man was ever held accountable for these crimes, and Vides Casanova was invited to the School of the Americas as a guest speaker in 1985.

Yet despite ample evidence of the involvement of security forces in human rights violations and a few victories for human rights activists, widespread impunity remains the norm throughout Latin America. Amnesty laws passed in the waning days of war and military rule, or enacted by unsteady civilian governments, continue to shield the guilty, and unreformed militaries protect their own behind a wall of secrecy, threats, and lies, claiming that national "reconciliation" depends upon burying the past. Human rights organizations and activists in the judiciary, however, have not given up their struggles to hold perpetrators accountable. Spanish judge Baltazar Garzón, for example, has played a leading role in bringing Latin American military officials, such as Augusto Pinochet, to justice. In addition, Garzón has requested that British authorities question former Secretary of State Henry Kissinger, who visited London in April 2002, about his knowledge of the international terrorist network known as

Operation Condor, after documents related to Condor, bearing Kissinger's signature, were declassified by the United States. Garzón has also waged a vigorous judicial campaign to root out al-Qaeda terrorist cells in Spain, where they are known to operate. The United States, however, hinders the fight against impunity by refusing to release documents that could shed light on human rights abuses around the world and the involvement of U.S. officials in them. It has also sent a strong message about the rights of the powerful by withdrawing from a treaty to create an International Criminal Court that will hear cases on war crimes and crimes against humanity. Bush administration officials fear that U.S. soldiers and diplomats might be prosecuted in this court, and, in an effort to undermine the court's power, the administration has pressured numerous countries to sign agreements in which they pledge not to hand over U.S. citizens to the court.

Victims and survivors claim, however, that justice and accountability must take place for societies to come to terms with brutal, authoritarian pasts. They say that remembering is important in order to forget (Hayner 2001). Recent events in Chile, sparked by the 1998 arrest of General Pinochet in London, bear out their assertions. After years of relative quiescence and concerted official attempts to suppress and reinterpret the significance of the Chilean dirty war, Pinochet's legal difficulties punctuated the aura of invulnerability that surrounded the former dictator. An intense national debate arose in Chile about his regime, the fate of the disappeared, and the persistence of impunity.

Impunity is an aspect of power that reinforces a highly unequal social order. Although it generally refers to the lack of accountability enjoyed by militaries, police forces, and paramilitary organizations, impunity can be usefully conceptualized in a more dynamic fashion as an aspect of power that is embedded in the process of social differentiation (Sider 2000) and that extends from the military and powerful civilian elites to the oppressive economic policies of international financial organizations. When experienced from above, impunity allows perpetrators to harm others without suffering consequences themselves, and when endured from below, it restricts the ability of people to limit violence—political, economic, and cultural—and hold perpetrators accountable. The impunity-fueled violence that has swept Latin America shattered the social bonds of those who survived the repression. It displaced people from their lands, drove them into unemployment, forced them to leave their homes for an uncertain life in exile, and deprived them of the friends and loved ones who supported them and gave meaning to their lives.

Men were targeted disproportionately, and women assumed a heavier burden in their absence. In the Guatemalan countryside, for example, peasant women had difficulty cultivating the land in the absence of male relatives whose labor they depended on (Green 1999), and female survivors from Central America to Argentina were pushed deeper into the labor force, obliged to work longer hours away from home to support their families at precisely the moment when children needed them most (Munczek 1996; Hollander 1997). Many victims and survivors continued to live in a state of fear, because informers, torturers, and local-level enforcers who collaborated with security forces continued to live in the same communities and neighborhoods with them. A sense of bad faith developed, as people did not know whom to trust. Because perpetrators often remained in positions of power, making demands for justice, accountability, and help in rebuilding shattered lives was difficult, if not impossible. The violent destruction of unions, cooperatives, and other grassroots organizations through which people made claims on the state, employers, and international agencies only aggravated the social fragmentation and economic vulnerability of survivors (e.g., Menchú 1984; Feitlowitz 1998). New forms of vigilante justice emerged in some areas where the state was unable, or unwilling, to hold perpetrators accountable.

The political violence, and the impunity that made it possible, thus undermined the ability of people to take care of themselves and generated new kinds of insecurities. In many countries, it preceded the enactment of free-market economic reforms in the 1980s and 1990s that mandated lower tariff barriers, cut social services, privatized public utilities, aggravated unemployment, and increased the gap between rich and poor. These reforms were demanded by the International Monetary Fund to facilitate the payment of large foreign debts incurred during the dictatorships, and to make it easier for multinational corporations to penetrate domestic markets and exploit the land and labor of ordinary Latin Americans. Reforms were generally implemented by civilian politicians through executive decree and exposed peasants and working people to greater manipulation by dominant groups. Illegal immigration to the United States and an influx of desperate peasant settlers to the coca-producing regions of the Andes were just some of the consequences. Common crime exploded, especially in Central America, where the distinctions between victims and victimizers became harder to discern in rural Guatemala (Godoy 2002). The consolidation of what became known as neoliberal capitalism required broad impunity for the powerful, and it depended on the maintenance of strong security forces to maintain "order" in the midst of increasing social decomposition and disorder that were themselves the result of the state's own policies. Order and disorder were thus closely connected in the state-sponsored, political and economic violence that plagued Latin America.

Despite the persistence of powerful, largely unreformed armies and the creation of new, militarized police forces, much about the security forces of the Americas remains invisible, hidden behind public relations campaigns, propaganda, and national security laws. All of this inhibits a full accounting of past and present actions, and it abets contemporary political violence in Latin America, such as in the Andean region. Thousands of peasant families have migrated to the coca-producing regions because there are no viable economic alternatives to coca growing, but they are labeled criminals and targeted by security forces—some of whom trained at the School of the Americas—who are noted for their abusive practices. These peasants understand confrontations with U.S.-trained security forces as attacks on their livelihood, not an international war on narcotics.

Ending military impunity has long been a goal of Latin American human rights organizations. Truth commissions convened in the aftermath of war and dictatorship have had only limited results, and activists continue to fight impunity for past abuses. For example, after more than two decades, the Argentine mothers and grandmothers of the disappeared continue to gather every week in the Plaza de Mayo in central Buenos Aires to demand that the military reveal the fate of their loved ones. Similarly, the Colombian human rights movement is trying to document violations over the last three decades and to build public support against impunity. Their efforts are complemented in the United States by a decade-long campaign to close the School of the Americas. If ending impunity requires an erosion of the power wielded by the militaries of the Americas, then understanding how they are trained, how they perceive the world, and how they confront powerless people in Latin America and critics in the United States is important. Using the tools of ethnography and archival sources, this book attempts to chip away at the facade of impunity by examining these processes.

The Research and the Book

In what follows, I relate a story of empire building in the Americas that turns on the violence of the state and depends on soldiers, agents of empire par excellence. It is a tale that I tell from the experiences and perspectives of three different groups: the U.S. and Latin American soldiers who pass through the School of the Americas and form part of the foundation of U.S. control in the hemisphere; the peasant coca growers of Colombia and Bolivia, who currently bear the brunt of state-sponsored violence in conflicts backed by the United States; and the U.S. activists who oppose the School of the Americas and the U.S. military's practices in Latin America. I have placed these groups within the same analytic framework because their struggles, as they take place "on the ground" at the SOA, in the jungles of Colombia and Bolivia, and on the streets of the United States, illustrate relationships and associations that are often denied, covered up, or ignored. By considering them together, we can begin to locate the fields of force in which the experiences, skills, and understandings developed in military training become transnational and tied to an imperial project that has consequences for people throughout the Americas. We can also better understand the claims and protests of anti-SOA activists in southern Georgia or peasant coca growers in the Andes, who advance different visions of the future of the Americas. Explicating these connections would be impossible if I were to maintain a focus solely on the School of the Americas. By including people who occupy different positions of power and privilege in the Americas, I explore what historian Julian Go describes as "the chains of empire" (2000).

I began by visiting the SOA in the late summer of 1999, just a few months after the U.S. House of Representatives voted to cut the School's funding, and I returned to the School on numerous occasions over the next three years. During these visits, I sat in on classes, interviewed students and officials, collected material from the library, and talked to people at Fort Benning, where the SOA is located, and in the adjacent town of Columbus, Georgia. Because of my frequent visits, some people at the SOA came to see me as a more serious listener than other civilian critics who visited the institution in the wake of a human rights controversy that erupted in 1996. This was perhaps why the commandant permitted me to accompany an officer class on a week-long tour of Washington, D.C., in the summer of 2000 and why he assisted me in setting up interviews with alumni in Latin America.

I had neither the time nor the resources to visit every Latin American country, so I decided to contact graduates from three countries: Colombia, Bolivia, and Honduras. All of these countries sent large numbers of trainees to the SOA. Colombia routed more to the School than any other Latin American country, and, along with Bolivia, it had done so consistently since the early days of the cold war. Honduras, on the other hand, was transformed into a U.S. battleship only during the Central American crisis of the 1980s, when it figured prominently in the Reagan administration's efforts to overthrow the Nicaraguan Sandinistas and to destroy guerrilla movements in Guatemala and El Salvador. During this period, Honduras dispatched hundreds of trainees to the SOA, but then slipped once again into the backwater of U.S. foreign policy and faded from the U.S. military's radar screen.

Although I did not make contact with all of my interviewees through Colonel Glen Weidner (see chapter 4), the majority of the interviews were arranged with his initial assistance. Weidner was himself an SOA graduate, and I chose to begin with a high-ranking U.S. official in the hope that he could open doors to Latin American students and alumni, as well as lower-ranking U.S. instructors and students who otherwise would have been hesitant to talk to me. This was in fact what happened.[10] Weidner gave me access to the classrooms of the School and vouched for me to various U.S. Military Group representatives stationed in Colombia, Honduras, and Bolivia. The latter, in turn, contacted local military commanders who drew on their personal networks of retired colleagues, former military school classmates, and active-duty officers. There were obvious drawbacks to this approach. Like most interviewees, mine told me what they wanted me to hear, but because the stakes were higher than in other kinds of interviews, their efforts to communicate a particular vision of reality were even more intense than what one usually expects. Few had anything critical to say about the School of the Americas. They exposed no secrets and never incriminated themselves, although a few pointed fingers at others. Many, in fact, probably did not have blood on their hands, since they were either connected to units that did not participate in human rights violations, or they were out of the country during periods of intense violence, but I was less sure about others.

They were all willing to tell me their version of their military careers and their experiences at the SOA, and most seemed happy to talk to an academic, who, they hoped, could write an authoritative account of the school. Sometimes I felt like a scribe dutifully noting the details of a revisionist history that obscured and silenced the past, and there were times, especially after a long day of interviews, when I could barely contain my anger and impatience with their dirty-war justifications and self-serving explanations. At other times, I was simply bored, but the worst moments came when after listening to seemingly endless justifications of the military's past and present mission and suppressing my own opinion, I felt nothing at all, just numbness. Thankfully, this was not always the case. Their accounts frequently illuminated aspects of training that I did not fully understand, and sometimes the men were unexpectedly revealing, such as when a question about human rights provoked their indignation, or when a casual remark connected to something that I had heard elsewhere.[11]

Like other people who have studied political violence, I was prepared not to like these men, and I had a hard time thinking about them as "ordinary" or "just like us" (Rosenberg 1991, 17; Robben 1995; Browning 1992). Yet some of them struck me as warm, engaging, and concerned about the welfare of others, and I never encountered the monsters I had imagined. One does not have to be a sociopath to do or to support terrible things. The problem lies less in "flawed personalities" than in our relationships to each other, and as Stanley Milgram (1974) concluded in his famous experiments, the most terrifying thought is not that violence can be inflicted on us, but that we—all of us—are capable of inflicting it on others through the ordinary, power-laden social relationships that structure our daily lives.[12] It is the rare person who can resist the pressures of conformity and authority to act in ways that are morally autonomous (see, e.g., Browning 1992). My job as an anthropologist was to listen and to try to grasp their understandings and perspectives. I was interested in how they described the experience of military training in the United States, the quality of relationships with U.S. officers, the nature of the opposition in their nominally democratic countries, and how they viewed human rights.[13] During the course of my conversations with these men, I did not reveal my political opinions to them, and, although I have tried to capture their views to the best of my ability, I suspect that most of my interviewees would neither agree with my analyses nor support my conclusions.

Interviewing peasant coca growers in the Chapare region of Bolivia and the department of Putumayo in southern Colombia presented a different series of problems. Because of the political violence and insecurity that currently engulfs these regions, especially Putumayo, doing the kind of in-depth observation and interviewing that is the hallmark of the anthropological method was impossible. Being a citizen of the United States did not help, either. It identified me with the government that waged war on local people through the forcible uprooting or aerial fumigation of their coca fields. Nevertheless, including peasant coca growers in this study was important, because they currently experience the invasive, militaristic policies of the United States more directly than any other group of people in the Americas. A book about military training that did not attempt to connect the activities of U.S. military schools to local-level contexts in Latin America would miss an important dimension. I benefited from the assistance of human rights workers and organizations who made it possible for me to travel—albeit briefly—to Chapare and Putumayo, meet with peasant families, local leaders, and, inadvertently, members of the security forces. These visits and interviews, as well as secondary documentation provided to me, were crucial in forming my understanding of state-sponsored violence at the local level and its consequences for ordinary people.

Negotiating the dynamics of power and the real threats of violence was much less problematic among the largely white, middle-class activists who participated in the movement to close the School of the Americas. These people included not only students who had studied at my university and, in at least a few cases, taken courses with me, but also many other people of various ages with whom I shared similar life experiences and political perspectives. Even though I identified with many of the goals of the movement, maintaining a critical distance about the campaign to close the SOA was important in order to appreciate the different perspectives and shifting dynamics within the movement, as well as the more ambivalent or critical feelings that some people experienced (Edelman 2001; Burdick 1998).

The chapters are organized in the following manner: Chapters 1 and 2 situate the School of the Americas in the context of the U.S. military, Georgia, and the campaign to shut it down. Chapter 1 examines the Army's current efforts to incorporate Latin American trainees into the cultural practices and understandings of empire through exposure to what it calls the "American way of life" in southern Georgia. The chapter considers the ways that different kinds of soldiers actually experience life in the United States and at the School of the Americas. Chapter 2 discusses how the SOA has struggled in recent years to refurbish its image in the wake of a human rights controversy that has enveloped the School. It considers how officials justify the mission of the SOA and how they have developed a public relations offensive to counter critics' claims that the institution is a "school of assassins."

Chapters 3 and 4 move beyond the halls of the SOA to Latin America. Chapter 3 examines the creation and growth of the SOA in the context of cold war Latin America. It argues that U.S. dominance in Latin America after World War II was based less on unilateralism than on the increasing collusion of Latin American military establishments with U.S. military power. In this process, the School of the Americas was an important center where these ties were cemented. The chapter explores how, in the context of shifting global realpolitik, the SOA shaped militaries from across the Americas into proxy forces under U.S. control and bought their cooperation by providing trainees with opportunities to participate in a cosmopolitan, modern world and to bask in the refracted glow of empire. Chapter 4 then takes up the particular experiences of one Bolivian trainee, who was first trained at the Bolivian military academy by instructors formed at the SOA, and then traveled to the SOA himself. It demonstrates how cold war militarization, and particularly the School of the Americas, opened pathways to power for male members of the lower middle class, even as it restricted nonmilitary educational and professional opportunities for them in their home countries. The chapter also shows how arming a segment of the middle class and providing it with channels of upward mobility aggravated racism and class-based forms of exclusion locally.

Chapters 5 and 6 return to Georgia and consider the experiences of the upwardly mobile officers of the SOA's flagship Command and General Staff Officers course. Chapter 5 analyzes how the SOA uses this class to train and build ties to Latin American militaries' future senior officer corps, and how these ties operate as students disperse and begin to circulate around the Americas. Through the manipulation of relationships with SOA graduates, the U.S. Army is able to gain better access to Latin American militaries, understand their internal dynamics, and exert control over them. Chapter 6 places this analysis in the context of the human rights training that students receive at the SOA and the ways in which student beliefs about human rights are shaped within their own militaries. It argues that the soa's human rights program is part of an ongoing argument with critics about the past, present, and future of the Americas, and that engaging human rights, but ignoring impunity, is an attempt by the soa to justify its continued existence without accounting for past practices or the behavior of some of its graduates.

Chapter 7 takes up the issue of impunity from a different vantage point. It explores how the dynamics of impunity shape the changing relationship between state-sponsored violence and deepening social fragmentation in two coca-growing regions of the South American Andes—the Colombian department of Putumayo and the Bolivian province of Chapare—where "waging war on drugs" provides justification for U.S. military intervention. Impunity-fueled violence destroys the social relationships that local people use to take care of themselves and each other. It also intensifies inequalities among people who can neither control the violence nor hold the perpetrators accountable. The disaffections and new forms of struggle that typically arise generate new calls for military training and intervention, as state agents seek to control the disorder that they constantly create.

Chapter 8 shifts back to the United States and examines a decade-long movement to close the School of the Americas and end military training for Latin American security forces. It traces the movement's roots to the Central American solidarity movement of the 1980s, discusses how it built an organizational structure, and explores the changing tactics used by activists and the military as they contend with each other. It then assesses the success and failures of the movement on the eve of the September 11, 2001, terrorist attacks.

Finally, the conclusion situates the preceding discussion within the new political and moral climate that has abruptly arisen in the wake of the September 11 terrorist attacks on the United States.

CHAPTER 1

Georgia Not on Their Minds

The first shafts of morning sunshine fall on the graceful contours of its red-tile roof and bathe the stucco walls in soft pink overtones. It appears that the stately building is blushing. Yet as the formidable Georgia sun continues its ascent, and the building falls under the intensifying rays of a late summer day, the walls turn a harsher, institutional beige. There is no shame here. This is the United States Army's School of the Americas (SOA), a military training center for Latin American men.

The SOA, or the Western Hemisphere Institute for Security Cooperation as the army now likes to call it, sits deep within Fort Benning, a sprawling army base on the southern edge of Columbus, Georgia. The institution bristles with imperialist symbolism. Emblazoned on the School's crest is the Spanish galleon on which Christopher Columbus conquered the Caribbean, and inscribed around the edge is the SOA motto, "One for all and all for one." This slogan comes from the Monroe Doctrine, James Monroe's message to Congress in 1823 articulating the nascent imperial ambitions of the United States by seeking the exclusion of its European rivals from the Americas. The words continue to describe the imperial unity that the School seeks to build among the contemporary militaries of the Americas.

The SOA is a relatively small part of Fort Benning, which spreads across 184,000 acres of a former cotton plantation that fed Columbus textile mills in the nineteenth century. The base is a place where thousands of people live, work, and train, and it is currently the engine that powers the local economy and sustains many of Columbus's 250,000 residents. Although its rolling, pine-covered hills serve as firing ranges and training grounds for the thousands of infantrymen and women who rotate through the facility every year, Fort Benning has many of the trappings of a small city. It contains living quarters, fast-food restaurants, banks, a library, movie theaters, a post office, a disco, schools, and a golf course. Those who live on the base are never far from work. They use the same facilities, send their young children to the same elementary school, and do at least some of their shopping at the tax-free post exchange, or PX. Active-duty service-men and women, civilian employees, and retired military personnel in Columbus receive approximately $950 million annually from the military in the form of wages, salaries, and pensions, and local businesspeople get millions more for services provided to the base.

The School of the Americas is located in a quiet area that seems less like a military base than a suburban neighborhood. On its serene lawns, where tall trees provide relief from the summer heat, blasts from the live-fire ranges and the din of training drills do not disrupt the tranquility. "River-side"—a restored plantation great house that is the official residence of Fort Benning's commanding general—is down the street. Its white pillars and manicured grounds recall the power and pretensions of the Old South's slave-owning aristocracy and symbolize, without a trace of irony, the vaunted position of its current occupant. The upscale homes of army colonels lie within walking distance on leafy residential streets. These two-story dwellings with white stucco walls, red-tile roofs, suburban-style lawns, and abundant shade stand in sharp contrast to the rows of monoto-nous, sun-baked barracks that house enlisted men on the other side of the base. The SOA's commandant and his family occupy one of them, near the eighteen-hole golf course, where, after many years of military service, they finally enjoy what the commandant refers to as "country club living."

Students, instructors, and administrators begin arriving at the SOA around eight o'clock in the morning. Some don heavily starched combat fatigues, stuffing their pants legs into imposing black boots. Others wear the Class "A" uniform—known as "greens"—a standard attire prescribed for daily use. All have close-cropped haircuts that disappear under a vari-ety of caps and berets. An occasional civilian administrative assistant or an English-language instructor, who is usually a woman, enters the building with the others. Wearing a printed skirt, a striped jacket, or a brightly colored blouse, these women stand out amid the monotony of earthen

Figure 1. School of the Americas. Photo by author.

tones worn by the men and the handful of military women who train at the institution.

The students have been awake since dawn, when the collective grunts and exertions of hundreds of people initiating physical training—"PT" in army parlance—bring Fort Benning to life. Joggers compete with early morning traffic on the streets, and open fields fill with soldiers going through the paces of morning calisthenics. Fat is anathema to the army. A bulging waistline or a double chin signifies moral lassitude and an unacceptable lack of combat readiness. For SOA students, the battle of the bulge starts every morning at 6:30 a.m., when they congregate on a field for a strenuous one-hour workout. After sweating through the exercise ritual and imposing order on their bodies, they shower, dress in the uniform appropriate for the day's activities, and head to the SOA, where they spend the rest of the day in field-based exercises and classroom lectures.

The SOA moved to Fort Benning from the Panama Canal Zone in 1984. Its trainees come from every Latin American country except Cuba, Haiti, Panama, and, until recently, Nicaragua, but their exact numbers and the countries represented at any particular moment reflect shifting U.S. political concerns in Latin America. Panama, for example, maintained a significant presence until tension between the United States and General Manuel Noriega led to the exclusion of Panamanian trainees in the late 1980s. Nicaragua also figured prominently at the SOA prior to the 1979 Sandinista revolution, which overthrew the Somoza dictatorship, dismantled the national guard, and prompted the United States to exclude the Sandinista-controlled military from the School until 2002. Nowadays, the majority of soldiers come from Mexico and the Andean countries of Bolivia, Colombia, and Peru, where the United States is waging war on the illegal narcotics traffic.

The School offers thirty-five courses on a range of topics to officers of various ranks, as well as a helicopter training program at Fort Rucker, Alabama. With the exception of the forty-eight-week Command and General Staff Officer course, most classes last from one week to four months. They include several geared to the requirements of cadets and noncommissioned officers that cover logistics, leadership, intelligence, and combined arms operations. Others focus on counterdrug and small unit operations, computer literacy and "information operations," formerly known as psychological operations, or simply PSYOP. The School has also developed a series of relatively new offerings, such as de-mining, civil-military relations, resource management, and human rights, in part to rebut the claims of critics who charge that it is a "school of assassins."

A precursor to the School of the Americas—the Latin American Ground School (LAGS)—was established in 1946 in the Panama Canal Zone, where the United States had trained Latin Americans at a variety of military bases since 1939. The Latin American Ground School centralized these activities and, after a subsequent reorganization and name change, became the School of the Americas in 1963. Its departure from Panama in 1984 coincided with the relocation of other U.S. military installations away from the Canal Zone, as the Panamanian government prepared to take control of the canal and the surrounding area.

Fort Benning was not an obvious choice for the School's new home, and substantial debate, intense lobbying, and backroom political wrangling preceded its selection. The army considered reopening the School on bases in Puerto Rico, Miami, or San Antonio because of the presence of large Hispanic communities. It also contemplated handing the institution over to Panama, but, in the end, a successful lobbying campaign by Georgia politicians and local businesspeople captured the School for Fort Benning. Sal Díaz-Versón Jr. and his sister, Elena Amos—anti-Castro

Cuban immigrants who made a fortune in the insurance industry—were at the forefront of this campaign, and their efforts merged with the desire of the Columbus Chamber of Commerce and Georgia congressional representatives, especially the powerful Senator Sam Nunn, to bring the SOA to Fort Benning.

Intense anticommunism and sharp business acumen drove the Cuban Americans' bid for the SOA. Like many Cuban immigrants of their generation, brother and sister shared a deep hatred of Cuba and its communist government. This hatred was nurtured by their father, Sal Díaz-Versón Sr., a right-wing journalist who dedicated his career, before and after the Cuban revolution, to denouncing communism. After the triumph of Fidel Castro and his revolutionaries in 1959, the Díaz-Versón family fled Havana for a life of exile in Miami. Sal Jr. was nine years old. He passed his formative years watching his family recover from its losses in Cuba and imbibing the bitterness of other refugees who, at least temporarily, had lost their wealth and the taken-for-granted privileges that accompanied it. Forty-one years later he described himself to me as "one of those hardliners that says Fidel has to go before anything changes." His older sister Elena shared these views. Prior to her death in 2000, she served for several years on the board of directors of the extremist Cuban American National Foundation, and she enjoyed important political ties with a number of conservative, fiercely anticommunist U.S. senators, including Jesse Helms, Orrin Hatch, and Strom Thurmond. The senators had been associates of her late husband, John Amos—the founder and chairman of the Columbus-based American Family Life Assurance Company (AFLAC).[1] Home-grown and Latin American anticommunism, coupled with capitalist self-interest, constituted the glue that bound together the Cuban immigrants and the conservative Southern politicians.

Elena and Sal Jr. believed that supporting the School of the Americas would prevent Cuban-style revolutions from happening elsewhere in Latin America. Mobilizing others to the cause assumed particular urgency for Sal and Elena in the early 1980s, when revolution appeared imminent in much of Central America. The Nicaraguan Sandinistas, who had overthrown the Somoza family dictatorship in 1979, represented an alternative to authoritarian regional regimes backed by the United States. The Sandinistas initiated a widely acclaimed literacy campaign modeled after the Cuban experience, and they decreed a popular agrarian reform that divested the Somozas and a clique of their associates of accumulated wealth in land. As the Sandinistas set out to remake Nicaragua in the 1980s, civil war intensified in Guatemala, where a brutal army counterinsurgency campaign left a trail of death and destruction in its wake. In neighboring El Salvador, guerrillas of the Farbundo Martí National Liberation Front (FMLN) seemed on the verge of wresting power from the military and the infamous "fourteen families" who ruled the country. Sal Jr. explained his sense of impending doom at the time and the importance of the SOA: "There was a cold war taking place in Latin America. The influence of Fidel and the Russians was real . . . They were the ones undermining democracy. El Salvador, Nicaragua—those were the key battlegrounds. [The SOA] is much less costly to the United States than having to send in U.S. troops . . . We [train] good military folks that can handle the problem internally. You do not need outside forces to come in."

Sal Jr. and Elena were not just driven ideologues. They were also astute businesspeople who understood the importance of promoting capitalism locally. The insurance company AFLAC had grown into an economic powerhouse that figured prominently in the Columbus economy and on the national scene, and Sal Jr., who was the company's president from 1978 to 1992, shared some of the responsibility for its success. He was also a leader in various business associations; for example, he had chaired the Atlanta Hispanic Chamber of Commerce and had served as president of the U.S.-Cuban Business Council, a national lobbying organization based in Washington, D.C. Perhaps most important, however, Sal Jr. and Elena appreciated the enormous economic potential of Fort Benning for Columbus. They believed that even the relatively minor contribution of the SOA and its $3.5 million annual budget would strengthen Fort Benning and bolster the bottom line of local businesses. As Sal Jr. explained: "The growth of Fort Benning helps the community. It has an economic impact—in car sales, in retail, in housing and so forth. So anytime we see the possibility of bringing another division or whatever to Fort Benning, we try to compete for it."

Yet there are reasons to question whether the impact of an enormous military installation on the local community is as beneficial as Díaz-Versón claims. In a study of Fayetteville, North Carolina—the home of Fort Bragg—Catherine Lutz documents a number of problems that plague military company towns. Small businesses, for example, are vulnerable to large fluctuations in the post population, and the army pays no property taxes because federal land is exempt. The absence of tax

dollars undermines the quality of public education and forces local people to subsidize the education of military children who attend public
schools. Furthermore, U.S. soldiers avoid state income tax because they
are allowed to maintain their home-state residency, and the overwhelmingly male nature of the military encourages a lively sex industry (Lutz
2001, 180–93). These criticisms of the army's behavior in Fayetteville apply to its conduct in Columbus as well, but Díaz-Versón did not address
them.

In 1983, the year before the School came to Georgia, Díaz-Versón
organized and chaired the School of the Americas Support Group in
which Elena played an active role, and he estimated that over the years he
spent between eighty thousand and one hundred thousand dollars of his
own money in soa-related activities. At its height, the Support Group
consisted of approximately forty prominent Columbus citizens who
spearheaded the campaign to capture the soa. Members knew that the
army brass was considering other locations with large Hispanic populations, and Columbus, at the time, could not boast of such a constituency.
Yet they remained undaunted and tailored a lobbying strategy that emphasized the greater "Americaness" of the Columbus metropolitan area.

They argued that soa students should not train in areas heavily influenced by Hispanics, because Latin American soldiers would feel too comfortable among their own kind and speak too much Spanish. To understand America, they argued, the trainees needed more access to "middle
Americans" and schools where their children could learn English. According to Díaz-Versón, "If you really want them to understand what
America is all about and how it really functions and meet the real people,
you have to put them somewhere like Fort Benning in Columbus."
Armed with this argument, he and others vowed to assist Latin American
officials and their families to assimilate into local life. They pledged to
invite students to their homes during holidays and to make them aware
of special observances, like Thanksgiving, that are not celebrated in Latin
America. They offered to connect individuals with health problems to
specialists who would waive high consultation fees, and they contacted
attorneys to help military visitors deal with traffic violations and other
legal problems.

The soa Support Group's rhetoric blended seamlessly with the right-
wing nationalism that flowered during the Reagan administration, and it
raised few eyebrows in the military; indeed, the army uses the same

reasoning today, as it struggles to justify the SOA and its presence in Georgia. Nowadays, old ideological chestnuts are retooled to address the human rights controversy that swirls around the institution. The military's public relations machinery churns out statements about the importance of sustaining democracy in the hemisphere by "engaging" Latin America militaries, and it paints a picture of the greater Columbus metropolitan area as an ideal locale for exposing Latin American soldiers to "American values," democracy, and "the American way of life" (read: white, middle-class, and heterosexual). Local military commanders express these views to anyone who will listen, but their cultural rhetoric defines, in subtle and not-so-subtle ways, what kinds of people count and by what measure.

It suggests that "real Americans" who practice the "American way of life" are still viewed by the military as white people who speak English, even though the army promotes itself as a racially progressive institution and officially acknowledges the equality of whites, blacks, and Hispanics. The army further claims that race consciousness is harmful and irrelevant to its operations. It downplays racial differences and exhorts new recruits to think of themselves as "green"—the color of uniforms. But despite the greater emphasis on tolerance and the real integration of the lower ranks of the army,[2] there is still a belief that divergence from a central set of values and practices is harmful to the United States and its national interests. Although official rhetoric dismisses race as a valid organizing criterion, the cultural superiority ascribed to those who partake in the American way of life ensures the same privileges that have always been based on racial discrimination.

A retired, white army major, who lives in suburban Columbus, describes the American way of life as "a good life." He explains many of the benefits that Columbus offers to the Latin Americans that are not always found in their home countries:

> Here in Columbus we are pretty much free of the concern that somebody is going to break into our home. We don't live behind fences and closed doors. We don't have security that has to patrol in front of our houses. We have good infrastructure, and we don't have to worry that the lights are going to go out when we have a party—I went to several parties in Panama where the lights did go out. You can trust the water sources, and we have good hospitals and ambulance services.

This description reflects a certain segment of suburban, middle-class existence in the United States, but it cannot be extended to poor, inner-city
neighborhoods or other precincts where the white middle class does not
tread. The major's emphasis on the good infrastructure is ironic, because
many public services in Latin America actually declined in the 1980s under
the free-market mania propelled by the United States. Privatization has
made them more expensive and less accessible than in the past, and this
has prompted widespread outrage in many countries.[3]

The paeans to the American way of life merely update nineteenth-
century racist notions about the "civilized" qualities of white Europeans
and Euro-Americans, rooting them in economic, cultural, and national
attributes rather than strictly biological ones. Such formulations reassert
the importance of extending these values to benighted peoples around the
world. This self-asserted superiority has justified almost any policy that
the "civilized" chose to enact on the inhabitants of the Third World, or on
the slaves, immigrant laborers, and indigenous peoples of the Americas.
Yet for all its extraordinary righteousness and certainty about American
virtue, the notion of an "American way of life" embodies a strong fear and
mistrust of foreigners. As Matthew Jacobson (2000) indicates, this kind of
nationalist thinking, which emerged in the nineteenth and early twentieth centuries, springs less from a sense of certainty about American
values than from a contempt for foreign peoples on whom U.S. citizens
depended as immigrant workers and overseas consumers.

The Latin American SOA trainees occupy a unique position within this
enduring nationalist vision. On the one hand, these Latin American visitors are closely linked to the most recent group of mistrusted foreigners
who have flooded the United States in search of work, and who allegedly
lack the values, sophistication, and civility that exposure to English and
the white, Southern middle class can presumably correct. On the other
hand, the United States military must depend on them to control the
political unrest and social disorder that threaten U.S. dominance in Latin
America, that send droves of impoverished, undocumented immigrants
across the border every year, and that arise to a considerable degree
because of the United States' own political and economic policies. The
heavy influx of Mexican and Central American migrant workers to Georgia and the rest of the United States in recent years is just one example of
how the legacy of U.S.-backed dirty wars and unbridled free-market pol-

icies have left the poor more insecure and unprotected and have touched off a new wave of population displacement and illegal migration.

The primarily middle-class provenance of the soa trainees sets them apart from the poor Mexican and Central American migrant workers who labor in the fields of southern Georgia, but they are still not immune to the U.S. Army's civilizing mission, which seeks to discipline them and to incorporate them into a U.S.-controlled, hemispheric military apparatus as junior partners and local-level enforcers. In a new twist to long-standing racist views of brutish Latin Americans, some soa officials account for the savage tactics used by Latin American armies against their enemies as a propensity that is somehow intrinsic to them. Brutality is, in other words, a character flaw that can be changed only when, in the words of one soa official, "the Latin Americans clean up their acts." They can do so, he believes, through exposure to the good citizens of Columbus and firsthand observation of a democratic system in action.

In this vision, Columbus stands out as a bastion of positive values, a font of democracy, and an uplifting way of life, but these images are historically false and destructive in important ways. In such a vision, U.S. participation in the making of such flawed Latin American militaries is entirely ignored; millions of dollars of military aid and decades of counterinsurgency training in U.S. military schools is believed to play no part in the creation of murderous security forces. Moreover, those with even a passing knowledge of United States history know of the terrorist violence waged against African Americans in Georgia by the Ku Klux Klan for the better part of the twentieth century. If they have read Carson McCullers's *The Heart Is a Lonely Hunter,* which portrays life in Columbus during the late 1930s, they appreciate the brutal racial hierarchies that are an intimate part of local history. They understand how this system of racial apartheid gave rise to the civil rights movement and spawned vicious police attacks on unarmed civil rights activists in the 1950s and 1960s. Indeed, the history of state and quasi-state violence in Georgia displays certain parallels to the dirty wars of Latin America. Both the Georgia Klan and various Latin American death squads acted with the complicity and participation of state security forces, and the gruesome tortures to which they subjected their victims were intended not only to eliminate dissidents but also to terrorize entire populations.

The symbols of the deep-seated "values" that spawned such brutality in Georgia have not disappeared. Confederate flags fly outside the homes of some whites, and the state flag that waves above state and federal buildings retains Confederate symbolism.[4] Public hostility has also greeted Mexican and Central American immigrant workers, who perform most of the agricultural labor in the state. Signs that read "English only" have reportedly sprung up in front of businesses around the state, and the conditions under which immigrants live and work raise questions about the newness of the so-called New South. Farmworker advocate Andrea Cruz, who advises immigrants in a multicounty region of southeast Georgia, insists that "if you say 'organize farm workers,' you might as well shut your doors and go hide. 'Cause that's a no-no." Not surprisingly, workers earn low wages and exposure to toxic pesticides is a constant problem. "The large farmers are sensitive to the issues," according to Cruz, "but we have smaller farmers who couldn't give a hoot. We've had . . . farm workers come into our office practically fainting. They went to work in a field that had just been sprayed the day before, and their eyes were dilated. Our migrant health program is so limited in this county and the benefits so poor." Cruz maintains that when the rhythms of agricultural labor intensify at harvest time and workers make slightly more money, the police set up roadblocks and put workers in jail on the slightest pretext. "As soon as they fork over the money, the police let them out," she explains. "It is like a weekly thing. The farm workers come here to make money but they end up leaving it. I believe that the police do it just to harass them."

A Latin American soldier, recently arrived to the SOA and Fort Benning, would have to search hard before he encountered large numbers of white, middle-class citizens who represent the "real Americans" of military imagery. African Americans, Latinos, and working-class whites represent overwhelmingly the middle and lower ranks of the U.S. Army, and they fill the barracks, discos, training grounds, and fast-food joints of the base. If the Latin American soldier stepped outside Fort Benning, he would see little change. Chatahoochee County on Fort Benning's southern perimeter is a poor, rural area with high rates of unemployment. On Fort Benning's northern border, an impoverished African American neighborhood adjoins the base and sprawls over much of Columbus's south side. Southgate—a low-income apartment complex of modest, two-story brick buildings—lies just outside the main entrance. Its residents labor at the bottom end of the service sector, where they fill jobs as health care aides, maintenance workers, retail clerks, auto mechanics, and secretaries. Some are employed at Fort Benning. The relationships of inequality that keep local

Figure 2. Georgia state flag, 1999. Photo by author.

residents in poorly paid, insecure jobs are eliminated from depictions of Columbus as a wellspring of American identity.

South Gate is only a block from Victory Drive, a multilane highway that separates Fort Benning from the bulk of the African American and Hispanic residents of Columbus. Victory Drive is the commercial hub of the south side and one of the first places where enlisted men from Fort Benning part with their earnings. It has more pawn shops, tattoo parlors, cheap hotels, seedy bars, and strip joints per mile than any other road in Columbus, and perhaps in the entire state of Georgia. The Columbus Pawn Shop is closest to the base and similar to many others. Stocked with guns, televisions, jewelry, cameras, and video recorders, its business turns on the first and fifteenth of every month, when servicemen and women are paid. A couple of miles to the west, military people mix in the Park 'N' Pawn Shop with black construction workers, single mothers, and other people from the surrounding neighborhood. The white co-owner, who describes herself as a "military brat," tells me that some of her best customers are army wives who buy jewelry. "The boys" from the School of the Americas, she adds, come in before they ship out to their home countries. "What do they buy?" I ask. "Guns and cameras," she replies.

Guns and cameras.

The conspicuous consumption of commodities is one of the national pastimes that SOA students participate in with enormous enthusiasm, and the fetishism of certain commodities, such as guns and cameras, is one of the simplest illustrations of how relationships of asymmetrical power are modeled internationally through a connection to things. Guns and cameras are manifestations of advanced "modern" states, and these consumer goods constitute valuable "trophies" for SOA students who return home after partaking of the American dream. Guns and cameras are part of the miraculous power of technology, which is an important measure of the perceived worth of societies, their relative power, and the value of individuals within them. Guns—especially the latest models available in the United States—symbolize male potency, and they convey real power to the men who wield them. They are indeed the most important tools in a global struggle to produce and accumulate commodities. Guns—and technology, more broadly—give their bearers an indisputable advantage in military conflicts that constantly occur between the self-proclaimed bearers of civilization and "modernity" and those allegedly "backward" peoples whose resources are appropriated in the name of "progress." Cameras are likewise the fruits of modernity, part of a wide range of technogadgetry that includes televisions, video recorders, computers, palm pilots, and so forth; these things make daily life more enjoyable and are simultaneously key symbols of a "modern," middle-class life that many SOA students and their families seek to maintain. Even though these commodities are more available in Latin America today than at any time in the past, they are not necessarily more accessible to the Latin American middle classes, from whose lower echelons most army officers originate.

By the end of the twentieth century, nearly two decades of debt-induced, fiscal austerity measures, economic restructuring, and the advent of free-market policies has taken a toll on the Third World. The United Nations characterized the 1980s as a "lost decade of development." The growth and consolidation of neoliberal capitalism in the 1980s and 1990s ushered in a period of severe economic turmoil for many countries, characterized by hyperinflation, currency devaluations, and political instability. State downsizing, a part of these changes, reduced an important source of jobs for the middle class. Lower tariff barriers resulted in a flood of globally produced commodities into Latin America and the collapse of local industries that could not compete with powerful transnational corporations. The allure of commodities tantalized people, who saw their purchasing power eroded by unemployment, inflation, and wage freezes as global media bombarded them with images of consumption. The symbols of modernity were at once closer to them and farther away.

Although these transformations affected specific Latin American countries at different times and with different degrees of intensity, their impact was widespread by the end of the twentieth century, and the foundations of middle-class life—job security, home ownership, and private education—were increasingly threatened (e.g., O'Dougherty 2002). Members of the armed forces were not immune. Budget cutbacks affected many Latin American armed forces, albeit not to the same degree as those state agencies charged with social services. Some officers found themselves struggling to maintain a middle-class lifestyle. Bolivian officers, for example, complained in the 1990s that low salaries and rising prices were forcing them to moonlight in other occupations and obliging their wives to work outside the home. The straitened circumstance, they claimed, were an affront to the dignity of the armed forces (Gill 2000, 115). Private security firms began to proliferate in Bolivia and elsewhere, as officers sought more lucrative employment opportunities (Castellanos 2000; Silverstein 2000).

For soldiers and their families who came to the SOA for a year, military training operated in part as a form of social welfare that enabled them to retain their grip on middle-class status at home. It not only promoted career mobility but also helped to consolidate their connections to a transnational vision of modernity. Their children could learn English in the United States. The possibility of acquiring cheap commodities was one of the central attractions of a sojourn in the United States for SOA students, who had certain perks and advantages that relatives and class peers in Latin America did not enjoy. Most, for example, received travel allowances in dollars, as well as their salaries. The dollars deposited in secure U.S. bank accounts shielded students from the wild currency fluctuations that frequently vexed the troubled economies of their home countries and the manipulation of savings accounts by governments desperately trying to manage crisis-plagued national budgets (e.g., O'Dougherty 2002). Latin American commanders used them as an incentive to motivate troops, and the U.S. military provided SOA students with access to Fort Bennings's PX, where taxes were waived and manufactured products were less expensive than in Latin America. A Colombian colonel explained: "The PX was a huge thing. Huge! Refrigerators, bicycles—whatever you wanted was there. Soldiers got all their stuff from the PX. It was a very good experience for them." The upscale Peach Tree Mall—Columbus's premier shopping venue—functioned as a weekend magnet, and seemingly incredible bargains in the pawn shops of Victory Drive emerged from the misfortune of local minorities, who, buffeted by the same economic forces that made life so difficult for many Latin Americans, had been forced to part with their own cherished consumer items. Participation in conspicuous consumption enabled trainees to maintain the appearance of class comfort and modernity in their home countries, while simultaneously reinforcing the status differences within their own cohorts.

In addition to their participation in conspicuous consumption, SOA trainees experience American life in other ways that are shaped by rank, nationality, and length of time at the school. Those men who come to the SOA for short courses of a few weeks leave their families in Latin America and typically reside in barracks on the base. These trainees, who constitute the majority of SOA students, do not receive attention from the SOA Support Group, nor do they participate in the School's Host Family Program, which is reserved for the officers of the School's prestigious Command and General Staff Officers course. When they are not training, these low-ranking students spend most of their time in the barracks and must take charge of their own introduction to U.S. society. They hang out in the discos and movie theaters on the base and visit the seedy local bars and strip joints on Victory Drive. On weekends, they may venture out to the mall to gawk at the window displays, and, before returning to Latin America, they typically pool their money, rent a car, and drive to Florida, where they make a pilgrimage to Disney's Magic Kingdom, which is considered almost obligatory by their peers and friends back home. Cameras are impotant tools for memorializing these quintessential "American" experiences.

The upwardly mobile officers of the SOA's flagship Command and General Staff Office course (CGS) have more complex experiences. These individuals, who are recognized as the future leaders of their respective militaries, come to the SOA with their families and remain for a year. Whether they live side-by-side in bleak quarters on the base, reside next to

African Americans in dwellings such as Southgate, or experience the all-white suburbs of north Columbus depends on their rank, race, and nationality. Rank determines to a considerable degree the size of an officer's living allowance, but nationality plays an important role as well. Military personnel from relatively well-to-do countries, such as Argentina prior to its recent economic collapse, are paid better than their counterparts from poor states, such as Bolivia, and they have a broader range of housing possibilities.

In Columbus, real estate values tend to increase as one travels north from Fort Benning, through the center city and into the suburbs, and locals orient themselves in reference to the exits on highway 285, which originates at Fort Benning, passes the east side of Columbus, and continues north until it merges with Interstate 85, the main highway to Atlanta. The most desirable housing and residential neighborhoods lie beyond exit 5, where whites' fear of crime fades along with the complexion of the residents. Yet every SOA student cannot aspire to live in this allegedly more "American" area. A white Argentine lieutenant colonel in the 2000 CGS course did so, but a sizable Bolivian contingent occupied cramped, one-story, adjoining dwellings in Battle Park, a residential enclave on Fort Benning. Their teenage children attended a predominantly black high school just off the base. The student body came as an initial surprise to some of the parents and students, who anticipated teenagers more closely resembling the white U.S. tourists who visit Bolivia and represent the quintessential gringo for most Bolivians.

The high school alarmed a Filipino-American trainee who was so upset by the lack of discipline that he sent his teenage son back to California to finish school. This officer was also having difficulty with his own studies at the SOA because he spoke very little Spanish. He explained that he was fluent in Tagalog—a language of the Philippines—but did not understand much Spanish. Yet a general sent him to the SOA, instead of a comparable course in English at Fort Levenworth, Kansas, because his last name was Gutiérrez. When I gave him an incredulous look, he insisted that it was true. Whether this individual was suffering from a case of mistaken identity, or simply disgruntled about his marginalization from the mainstream of the army, he clearly felt that he had been a victim of racism within the U.S. military.

Students experience the SOA in ways that are shaped by their own backgrounds and histories, and in ways that are far more complex than suggested by the military's propagandists. The military understands the inequalities among students, and it knows that trainees are well aware of the living conditions enjoyed by many United States military officials who serve in Latin America. These U.S. officials typically reside in upscale neighborhoods and enjoy a higher standard of living than in the United States. Yet the soa downplays these distinctions in practice. Hemispheric unity and equality among nation-states is the oft-stated message and the underlying theme of an unending series of social events.

Like the military brass in general, soa officials are energetic socializers, and they take their protocol and ceremonial obligations very seriously. Late afternoon gatherings to commemorate national holidays with fes- tive toasts and speeches are constant occurrences. So, too, are cultural events that feature the food and folklore of different countries. On these occasions, U.S. and Latin American soldiers celebrate imaginary "national cultures" through potent symbols and invented traditions, and they praise the armed forces for defending the lives and values of "a people." soa officials preside over these ceremonies like benevolent patrons who must occasionally intervene to contain the outbreak of tensions that disrupt the desired harmony.

Such was the case during Hispanic Heritage month in 1999, when soa students erected a series of displays at a local middle school. The purpose of the exhibit was to "promote sharing and understanding among cul- tures," but the Ecuadoreans disrupted the goodwill by displaying a map of Ecuador that depicted a disputed portion of neighboring Peru as if it really belonged to Ecuador. The map was like many sold in bookstores through- out Ecuador, but it provoked the nationalistic ire of Peruvian students, who marched back to the commandant's office to complain. After giving the matter some thought, the commandant decided to issue an "exhorta- tion" to all students. Without pointing a finger at anyone, it stated that politics was not to spoil the festivities. Afterward, according to the com- mandant, "the map came down."

In addition to these feats of social engineering, the soa incorporates Latin Americans into its bureaucratic apparatus. One Latin American always serves a two-year term as subcommandant, and the selection of a candidate moves in a deftly orchestrated rotation that takes care not to favor one country more than another. Latin American instructors also teach at the institution. They are typically chosen from the ranks of the cgs students who are asked at the end of their studies to remain at the soa for an additional year. They work side-by-side with U.S. trainers who are often Latinos and speak Spanish as their first language.

U.S. Latinos—instructors and students—find themselves at the nexus of a series of tensions and contradictions that shape the United States' efforts to build an imperial military apparatus and to make a particular vision of the American way of life transnational. Latin American students often prefer these instructors to their Euro-American counterparts. They feel affinity with the Latinos and believe that these individuals are more tolerant and understanding than the Euro-Americans. Yet at the same time, they commonly perceive the U.S. Latinos as "uppity" and chide them for putting on airs in an effort to separate themselves from their Latin American peers.

One former student from Peru commented that the Latino trainees had no trouble speaking Spanish in the company of monolingual Spanish speakers, but as soon as a Euro-American joined the conversation, the Latinos suddenly began to search for Spanish vocabulary, which they seemed to have suddenly forgotten. A Bolivian major painted a darker picture of people whom he described as "cynics." In his estimation, cynics were all those Latinos who acquired U.S. citizenship and then proceeded to "become more gringo than the gringos" (asumen los valores más fuertes que los mismos gringos). He explained: "When we have class discussions about strategy, for example, . . . the Latino North American will say something like 'but why are we making problems for ourselves. Let's just send them two more bombs.' Do you understand? They have this power thing stronger than the gringos—the arrogance of power, the conceit of power. I call them cynics because of a kind of cultural distortion that is so strong that they acquire values that are much stronger than those of the gringos." The major then went on to assure me that "the gringo-gringo [Euro-American] prefers to deal with Latin Americans more than Latino North Americans."

For the U.S. military, Latinos present a difficult conundrum. On the one hand, their cultural backgrounds and Spanish proficiency make them ideal instructors for the School of the Americas. On the other hand, their physical being and class backgrounds make them dubious representatives of what the School wants to portray as the American way of life. Furthermore, their presence tweaks national anxieties about "foreigners within." The remarks of a Euro-American, foreign-area intelligence officer speak to these tensions.

Born in a small town in rural Pennsylvania, the man became a Latin American specialist through a foreign-area training program offered by the army. The program enabled him to study Spanish and provided him with an opportunity to live in Latin America as part of his training. By the time I met him, he was based in Honduras. He had achieved a high degree of fluency in Spanish, ascended to the rank of major, and married a Costa Rican woman. I asked him why the army needed Euro-Americans like himself to represent it in Latin America. The major replied that his ability to speak Spanish was only part of his value. He insisted that he also represented the United States in a way that a Latino who only spoke Spanish could not. He explained that a Puerto Rican who spoke *jíbaro* [hick] Spanish, wrapped gold chains around his neck, and wore his shirt open to the naval was not the kind of person the United States wanted as a representative, and, he insisted, Latin American commanders would themselves be offended by such an individual.[5]

A former soa commandant, who ran the institution in the late 1980s, believed that the U.S. Army used the School as a "dumping ground" for unqualified Puerto Rican instructors. This practice, he believed, distorted the perceptions that Latin American trainees formed of the U.S. military. "Sometimes," he said, "the problems were that [instructors] were native speakers of Spanish. We wanted the School to be representative of the U.S. Army and not, for example, of Puerto Rico. It had been the tendency to fill the [instructor] slots with native speakers. Where do most of the native speakers in the army come from? They are mostly Puerto Rican."

Puerto Ricans were also heavily represented among the U.S. officers of the soa's Command and General Staff Officers course, and their presence and alleged lack of technical expertise made some of their Euro-American peers uncomfortable. A 1992 Euro-American graduate from the course complained in an article in *Military Review* that the School accepted too many Puerto Ricans into the class and used them as "filler" because of their Spanish language abilities. He urged the institution to select U.S. students on the basis of career orientation, rather than ethnic back-ground, to "better assure the professional contribution of U.S. officers to non-U.S. students." He went on to complain that eleven of the seventeen U.S. officers in his class were Puerto Rican and that only one of them was a Latin American specialist (Demarest 1994, 49, 51).

Despite the soa's official motto—"One for all and all for one"—race, ethnicity, class, and nationality are important organizing principles at the

SOA, even though at times they were officially denied. They shape the differential worth of the men who rotate through the institution, make the hierarchical rigidities of rank more contingent, and mold the ways in which diverse soldiers are incorporated into, or excluded from, a vision of imperial culture defined as the American way of life. SOA officials hope that, through direct experiences and contact with a carefully selected slice of America, the Latin American students will leave with a deeper understanding of "how America operates and does business" and return home willing to support U.S. policies in their own countries and collaborate with U.S. military officials.

In addition to their efforts to secure the cooperation of Latin American trainees, SOA officials remain resolute in their efforts to spin a story to the U.S. public about the soldiers and the School that emphasizes the uplifting benefits of exposure to the American way of life for armed forces with a seemingly natural propensity for brutality and undemocratic practices. The story paints a wholesome picture of a military school that trains people to kill and yet has been seemingly unconnected to the violence that has wracked Latin America. The story serves as the guiding myth used both to sell the School to a skeptical domestic audience and to incorporate Latin Americans into the practices and visions of imperial America. The military is particularly eager to relate this story to journalists, academics, and human rights activists.

⊕

CHAPTER 2

De-Mining Humanitarianism

"I am not a torturer," asserted Lieutenant Colonel George Ruff, "and I resent one of our representatives making these accusations on the floor of the House of Representatives." It was late September 1999, and Ruff was referring to Congressman Joseph Moakley's biting reference to the institution as a "toxic waste dump." The remark made Ruff and other SOA officials livid and intensified their outrage about a recent vote in the House of Representatives to cut the SOA's federal funding. On July 30, members of the House voted 230-197 to cut two million dollars from the School's budget, a move that would force the institution to close. Although a House-Senate conference committee overturned the vote weeks later by a close 8-7 margin, disturbing questions about the SOA graduates, their human rights record, and the School's training practices had been pushed into mainstream political debate by a social movement dedicated to closing the School. The movement had effectively ended the School's public anonymity and forced a congressional debate, and it worried many in the institution who no longer took the School's existence for granted.

Yet even as men like George Ruff defended their damaged reputations, the School of the Americas continued to work with the unreformed militaries of Latin America. Ruff himself was a career intelligence officer. He had just returned from three months in Columbia, where he was part of a mobile training team that instructed the first of three Colombian counter-narcotics battalions created by the United States. Mobile training teams sent to Colombia from the School of the Americas in 1999 instructed 555 Colombian soldiers, a figure that represented more than half the total number of in-resident trainees at the School that year.[1] Although the Special Forces carried out most of the training, "our mission," according to Ruff, "was with some of the senior people—the leadership," even though the Colombian military had one of the worst human rights records in the world.

soa officials insisted that "engaging" the armed forces was the best way to ensure that democracy would prevail in Colombia. They employed the verb "to engage" in the general sense of occupying "the attention or efforts of a person or persons," rather than the strictly military meaning of entering "into conflict with an enemy." This usage implied that by training Latin American security forces, the United States could curb the tendency of soldiers to commit human rights abuses. No training presumably led to even more brutality. Such claims, however, were hotly contested by a growing number of soa critics who understood "engagement" as just another military euphemism for training killers. Although officials had assumed a business-as-usual attitude since the vote in the House of Representatives, an aura of uncertainty and impending doom enveloped the soa, and the urgency of mounting a public relations campaign to counter the charges against the institution had intruded on the endless bureaucratic tasks of running a military training school.

I first encountered the soa amid this atmosphere of crisis. The first person I met was Nicholas Britto, the newly hired public affairs officer. Britto met me in his office, after I signed in at the front desk and climbed a wide staircase under a beautiful stained-glass window to the second floor. A short Cuban American with dark, wavy hair, Britto had moved recently to Columbus from San Diego. He was guardedly optimistic about his job, which had assumed more importance as pressure on the school mounted, and he seemed enthusiastic about living in Columbus. Local real estate prices were substantially below those in San Diego, and he had taken advantage of them to purchase a home. Britto had already visited several Latin American countries, and his office walls were decorated with mementos from these visits. A woven tapestry with a llama motif covered one wall, and he pointed proudly to a glass plaque on another wall, a gift from the Guatemalan military.

Like everyone I met that first day, Britto assumed that I was an anti-soa, human rights activist intent on unearthing dirt about the institution, but a broad smile never left his face during our conversation, and he exuded a cloying eagerness to address my concerns. The School of the

Americas is a totally open, educational institution, he told me. You can go anywhere in the building. If you want to sit in on the classes, that can be arranged. Torture is not taught here. These words would be repeated incessantly to me for the rest of the day by a variety of people.

The SOA's public relations strategy had entered a new phase. An inaccessible institution had seemingly opened itself to critics, and formerly reticent officials had been replaced by others who were more media savvy and adept at dealing with the human rights controversy. The military knew that it was losing the public relations war. The new "openness" constituted part of a last-ditch attempt to save the School and burnish its tarnished image by constructing a wholesome picture of its activities. One of the SOA's weapons in the public relations battle was a barrage of deafening doublespeak that blurred the lines between truth and fiction and sometimes obscured them completely. The first casualty was the English language, as officials manipulated a variety of stock phrases and clichés to describe military training as an exercise in democracy, engagement, and the promotion of human rights. The second victim was any perspective that deviated from the military's definition of reality.

War Is Peace

The School of the Americas's course catalog featured an array of anodyne course titles like Civil-Military Operations, Humanitarian De-Mining, Peace Operations, and Democratic Sustainment, which the army had created in recent years to mollify critics. These titles attempted to portray the School as an altruistic, civilian-friendly institution that was in tune with the concerns of the post–cold war world, but beneath this lexicon of benevolence lurked a darker reality. Civil-Military Operations was a case in point.

According to the catalog, Civil-Military Operations provided instruction in "military civic action, the proper role of the military in support of civilian authority, civil defense, disaster relief, and CMO [Civil Military Operations] support to counterdrug activities" (USARSA 2000, 66). Yet it was merely a new label on an old practice. So-called civic-action training had long been provided by the SOA as a tactic for fighting guerrilla movements. The civic-action initiatives targeted civilians thought to support insurgencies and consisted of a variety of local development projects that included the construction of schools and health clinics, vaccination campaigns, well digging, road building, and so forth, but they were not really about improving the quality of life for the poor. Military strategists designed the schemes to win the "hearts and minds" of local people and to support direct military objectives, such as constructing a road to provide access to conflicted areas. Civic-action projects were intrusive and manipulative efforts to put a benevolent face on the violence and destructiveness of counterinsurgency warfare, and, according to McClintock, "even when pursued in good faith, [they were] sabotaged by the ethos of 'anything goes' that dominated [counterinsurgency's] strictly military side" (1991, 44).

In Guatemala, for example, the military introduced a civic-action program in the late 1960s that accompanied massive government repression in the highlands and led to the deaths of five to ten thousand peasants. Modeled after the U.S. experience in Vietnam and funded by the U.S. Agency of International Development, the civic-action program, called "Operation Honesty," constituted the soft side of the military's efforts to pacify local people (Schirmer 1998, 36),and it included graduates of the SOA among its operatives. Between 1966 and 1968, the Guatemalan army sent five people, including three officers, to the School of the Americas to learn about civic action and "civil-military operations,"[2] and they almost certainly had personal contact with U.S. instructors fresh from the battlefields of Vietnam. Yet it should come as no surprise that the program failed to convince peasants of the Guatemalan military's good intentions, and a more hard-line sector of the office corps began to advocate "100 percent" brute force against alleged civilian collaborators in later years (Schirmer 1998, 36–37). This strategy reflected a similar transition in U.S. approaches that changed from efforts to win "hearts and minds" in the early 1960s to "coercive counterinsurgency," that is, unrestrained power, at the end of the decade (Robins 2003).

Humanitarian De-Mining attempted to airbrush the SOA's rough edges and highlight the institution's civic-mindedness. According to the catalog, the course taught mine, booby trap, and minefield detection techniques, demolition, and countermine methods (USARSA 2000, 82), and SOA officials rarely missed an opportunity to discuss the class with civilian visitors. As I became familiar with the School, the commandant suggested to me on a number of occasions that I attend a session, but I found the mawkish title to be such an obvious public relations gimmick that I steered clear of the class for a long time. Only after further urging by the commandant did I finally decide to join the students one afternoon.

I met up with Puerto Rican instructor Jesús Hernández and his trainees in a classroom where dozens of mines were laid out on long tables. Some were constructed to be set off by the metal in a soldier's boot; others detonated from the vibrations of a tank rumbling over the ground; and a tripwire discharged the "Bouncing Betties," which shot up six feet into the air before they exploded. The mines ranged from very small, antipersonnel devices, designed to blow the legs off anyone who stepped on them, to state-of-the-art, antitank contraptions produced by an Italian firm that, according to Hernández, also made lipstick in another part of its factory. The Italians sold these mines to Iraq, which used them against the United States in the first Gulf War, and Hernández, who served in the war, told me that he had not previously seen the devices when the army sent him out into the desert to deactivate them. Keeping up with the latest technology and knowing who was using it required a lot of work, he explained, because arms manufacturers were reticent about exposing the names of their clients, and few controls on the sale and purchase of mines existed. Hernández relied on U.S. intelligence operatives to pass on information about new mines to him, and, at the School of the Americas, he did his best to familiarize Latin American trainees with at least some of the wide spectrum of mines available on the world market.

From the classroom, Hernández and I drove out to the training field. It was an intensely hot summer day. The temperature hovered in the nineties, and the air was heavy with humidity. The air-conditioning in his car was barely functioning, so we drove with the windows down, and Hernández, who was craving a cigarette, lit up and smoked one after another as we continued. Waves of heat shimmered above the road that wound over the rolling, pine-covered hills of the base. When we pulled up to the range a few minutes later, the students had already arrived. Some were peering through the heat and hazy sunshine with binoculars, trying to locate mines placed near rotting tree stumps, in open fields, and in vegetation dappled with sun and shade. Even though signs with large numbers indicated their approximate locations, it was extremely difficult to see the devices.

Hernández explained that different kinds of terrain posed distinct challenges to soldiers sent to locate minefields. In the desert, he said, windblown sand constantly changed the contour of the land, but the dense vegetation and green monotony of a jungle presented other problems. A demonstration minefield a few feet away combined some of these challenges for the students. Hernández led me over to the field where a series of antitank devices lay buried at five-foot intervals. The mines had been in the ground for several months, and students could observe the effects of time and exposure on them. Some of the apparatuses were still invisible, but shifting soil had exposed others almost completely. The vegetation on top of still others had died, and dry brown spots distinguished them from the surrounding grass-covered field.

Hernández told me that he does not teach trainees how to activate mines, nor does he tell them how to deactivate them. The course focused exclusively on locating the devices, identifying them, and destroying them *in situ*. Given the enormously destructive potential of mines, and especially their capacity to kill and maim civilians years after the end of war, this seemed a worthy endeavor. Mines had been used in Central America, and warring groups used them in Colombia. In addition, some of the students wanted to use the training to participate in United Nations–sponsored missions to other parts of the world. Taking issue with a course that taught students how to destroy such deadly devices was difficult; despite its maudlin title, Humanitarian De-Mining seemed a valuable course.

Yet for all its apparent good intentions, the class was not necessarily as civic-minded as its title suggested. It exposed students to a quantity and variety of mines that far surpassed what they were likely to encounter in their own countries, and Hernández had commented earlier that the possibilities for using the devices were limited only by one's imagination. Even though Hernández taught students only how to identify and destroy the mines, placing and activating them was a much less complicated affair. Who could be certain what these students would do with the knowledge acquired at the SOA? And would SOA officials assume any responsibility if one of the students used the training inappropriately? But these concerns were not part of the compassionate image of the School that the course was designed to project. Humanitarian De-Mining was also not one of the SOA's marquee classes when it came to attracting Latin American soldiers. Like other courses marketed for their public relations appeal, it suffered from low enrollment; an average of only fourteen soldiers took the class each year between 1997 and 2000. The more traditional combat-oriented courses still captured the lion's share of the SOA trainees, and they remained the centerpiece of the School's military training program.[3]

Military Intelligence was one such class, but the course had been substantially revised after the public revelation in 1996 of objectionable training materials used in it. Known as the "Torture Manuals," these texts advocated the use of fear, beatings, the payments of bounties for enemy dead, false imprisonment, executions, and truth serum as methods of recruiting and controlling intelligence sources. They referred to extortion as a method of interrogation and appeared to advocate execution, or "neutralization," of enemies. Used in the course between 1989 and 1991, the manuals were also distributed by Special Forces Mobile Training teams to military personnel and intelligence schools in Colombia, Ecuador, El Salvador, Guatemala, and Peru. They emerged from the 1960s Army Foreign Intelligence Assistance Program, or "Project X," which supplied training materials to allied militaries around the world,[4] and they were brought to the School from the army's intelligence training center in Fort Huachuca, Arizona. The "Torture Manuals," like counterinsurgency doctrine in general, served as what McClintock described as "a kind of off-the-shelf, anti-subversive software for use anywhere and anytime" (1991, 134), but the outcry that followed their public disclosure sparked increased opposition to the SOA in the House of Representatives and among the general public. Consequently, the course underwent a major facelift and shed its sinister visage by the time of my visit. A special desk awaited visitors. A syllabus lay open on it, as well as a list of the students, showing their current grades and class standing.

The young instructor from upstate New York, Sergeant Ziska, invited me to sign a visitors' book. Ziska had worked with the Special Forces and participated in a peacekeeping mission in Haiti. He explained to me that Military Intelligence was a basic course for lieutenants who were recent graduates of military academies across Latin America, but higher-ranking officers often attended too. His current students included lieutenants, two captains, three majors, and one lieutenant colonel. All outranked Ziska, who noted that sometimes his trainees claimed to know more about intelligence than he did, and they used their superior rank to browbeat him, but Ziska insisted that U.S. privates usually understood intelligence better than Latin American officers. The commanders, he explained, relied less on electronic forms of intelligence gathering preferred by the United States than on human intelligence collected from sources by military operatives. Consequently, they selected officials to fill "S-2" (i.e., intelligence) positions based on personal connections and loyalty, and such people, according to Ziska, had very little training and tended to "shoot from the hip." He mentioned a former Salvadoran student who shared his interrogation methods with the class. The student explained how he blindfolded captives, took them up in a helicopter to disorient them, and then, with the helicopter hovering just above the ground, threw them out. He presented the procedure as a useful method for terrifying prisoners so that they would provide information. Ziska emphasized that, although U.S. soldiers used similar measures in Vietnam, the U.S. Army no longer advocated these tactics. Whether or not this was the truth, officials at the School of the Americas had in fact struggled mightily to cleanse Military Intelligence of any stain of impropriety.

The current commander of the soa's public relations campaign was its commandant—Colonel Glen Weidner, an articulate, West Point graduate who devoted much of his time to spreading the army's version of the truth about the institution. On my first day at the soa, I was quite surprised to discover that the public affairs officer had arranged an appointment for me to meet Weidner. At promptly ten o'clock, he ushered me into the commandant's office. During our three-hour conversation, Weidner was interrupted only once by his secretary, and he did not take any phone calls. This behavior left me perplexed. I wondered what the man did all day and how he could afford to take so much time to talk with me. But the long meeting was not an aberration. Over the next year and a half in which I became acquainted with Colonel Weidner, we would have more lengthy discussions, and I discovered that others in the media and academe received similar treatment. The military was intent on controlling the soa story, and, for Weidner, as for his two predecessors, managing the negative publicity and the protest movement was his biggest challenge. It was a task that interfered with the more typical ceremonial and managerial aspects of his job, but, according to a local journalist who had covered the soa for years, Weidner was well suited for the assignment because he was more accessible, politically astute, and better able to address the broader issues than previous commandants.

At our first meeting, Weidner wore combat fatigues, and several years at desk jobs had taken only a slight toll on his physique. After exchanging pleasantries for a few minutes, he sat in front of his computer and called up charts, tables, and graphs that debunked what he referred to as the school's "Black Legend," a myth propagated by "leftist missionaries linked to liberation theology." He insisted that the so-called Black Legend extended beyond the news media. It had penetrated popular culture and surfaced recently in an episode of the *X-Files*, a science-fiction television series that featured an evil alien with the SOA in its background. As he clicked through a monotonous series of multicolored pie charts and bar graphs, it became apparent that Weidner had given this presentation on other occasions. Sensing its tedium, he joked that he did not want to cause my "death by slides" but then continued relentlessly.

When he finished, we launched into a wide-ranging discussion about the school, Latin America, and the protest movement. The colonel spoke passionately in defense of the SOA and against an array of dissenters who, he felt, distorted the School's mission. The main target of his ire was Father Roy Bourgois, the Catholic priest who built the movement against the SOA and served as its most charismatic leader and spokesperson. Weidner also had little patience for the Vietnam veterans who had joined the anti-SOA campaign. "They're soldiers," he spat in disgust, implying that they were somehow traitors for opposing the SOA. At one point, he picked up a copy of the *Report on the Americas*, a left-leaning publication produced by the North American Congress on Latin America (NACLA), that lay on his desk and asked if I were familiar with it. The NACLA *Report* was one way that SOA officials kept tabs on what the U.S. left was thinking about Latin America. Its editors had angered Weidner by denying an SOA request to reprint an article about the legal difficulties of Chilean General Augusto Pinochet, who was arrested in London in 1998. Weidner characterized the organization and the editorial decision as "Stalinist."[5]

Weidner apologized several times for his vehemence about the human rights controversy that surrounded the School. The topic "makes me very emotional," he said, as his jaw tightened. Weidner's world was the military, and a strong institutional pride bound him to it. The son of an army officer, he grew up on military bases and then studied at West Point before beginning a long series of assignments in Europe, the United States, and Latin America. Along the way, he graduated from the school's Command and General Staff Officer course in 1986 and served on the faculty in 1987. When his two sons came of age, they, too, followed the footsteps of their father and grandfather into the army.

Acknowledging that "a few bad apples" from Latin America had attended the School of the Americas, Weidner insisted that these individuals were never taught torture techniques, and that their crimes represented the unconscionable acts of a few "rogue actors," not the teachings of the SOA or the policies of terrorist states. He maintained that some graduates, who stood accused of human rights violations, had only taken short courses on benign topics, such as auto maintenance, and had trained at the School years before their alleged crimes took place. It was unconscionable, he argued, for critics to point fingers at the School and claim that it *caused* these men to commit crimes. In a rationalization of the SOA that I would hear from others, Weidner pointed out that the Unibomber went to Harvard. Does that mean, he asked rhetorically, that Harvard caused him to kill people? Does that mean that Harvard should be shut down? Weidner—and others at the SOA—thus did not deny the reality of human rights violations, but his argument treated a prominent university and a military school as comparable institutions. Harvard, however, did not teach combat skills to Latin American soldiers; moreover, the United States government had used its military apparatus—including the SOA— to support Latin American armed forces with bad human rights records for decades. Yet if one objected to his confused logic, Weidner dismissed the critique as antimilitary, and thus unacceptable.

For Glen Weidner, the SOA represented "a redemptive effort for the Latin Americans, given the historical flaws in Latin American militaries." It was a place where virtuous U.S. instructors could impart democratic values to unruly Latin Americans who had a historical proclivity to violence, and who were seemingly unable to control their behavior. In Weidner's view, these values represented the essence of America and its military. I became more intrigued as I thought about his comments. Were his references to redemption and democratic values only public relations bromides that obscured the school's true mission—training soldiers to kill—in a warm, fuzzy glow? Or were they more than a smoke screen? To what extent, I wondered, did Weidner—the career military man—believe these platitudes? He need only talk to his colleagues and chat with appreciative SOA students to feel assured that he was not alone in his views, and many other people in Columbus, with its large population of retired military officials, would probably concur. Questioning the clichés might pose an intolerable threat to the world that had nurtured and shielded Weidner for his entire life.

Weidner piqued my interest because he did not fit my idea of an SOA official, whom I imagined as unidimensional, humorless, obsessed with order, and entrapped in a Manichean view of the world that excluded any appreciation of nuance or ambiguity. My image emerged from occasional, superficial encounters with U.S. military personnel in Latin America, when I visited an embassy to renew a visa, or saw them in public places. It was also shaped by statements that they made to the media in the United States and by my long awareness of the sustained support provided by the United States to Latin American dictators and the militaries that spawned them. I knew, too, that U.S. Army officials trained Latin Americans at the SOA, and that some of these trainees employed torture, extrajudicial executions, and campaigns of terror in their dirty wars against alleged subversives.[6] My view of the typical SOA official was shared, with certain variations, by many other people who were knowledgeable about Latin America and concerned about the precarious state of human rights there. Yet whatever the reality of the image, it was nevertheless a stereotype that resembled in its dehumanization the stereotypes of Latin American peasants, workers, students, human rights activists—indeed, any critic of the status quo—propagated by the United States and its regional military allies. It was therefore with a certain irony that SOA officials found themselves struggling against a process of degradation, as they sought to defend themselves against charges that labeled them "assassins."

Although I anticipated Weidner's well-crafted defense of the SOA, I did not expect to encounter a man who devoted so much time to civilian visitors, who seemed thoughtful and attentive to the views of others, and who opened the institution to me. As we became better acquainted, there were also times when Weidner went out of his way to assist me. On one occasion before the annual anti-SOA protest, he lobbied unsuccessfully a Fort Benning bureaucrat to give me a press pass so that I could cross into Fort Benning with the demonstrators and avoid arrest. His efforts on my behalf were always linked to his desire to project a particular image of the SOA and to control me and my understanding of the institution, but he struck me nevertheless as a decent man, who did his duty as he understood it. This view was also shared by others who knew him.

Weidner's abiding loyalty to the military made it difficult to dismiss his cliché-filled defense of the SOA as simple dissembling. The sociologist Zygmunt Bauman notes that in a hierarchical, bureaucratic institution, the concept of loyalty points to a higher authority as the ultimate source of moral direction, and means doing one's duty in accordance with the rules laid down by that authority. An individual is prompted to "measure his own righteousness by the precision with which he obeys the organizational rules and his dedication to the task as defined by the superiors" (1989, 160). Personal behavior thus becomes focused on job performance and discipline in a setting where conformity is valued more than independent thinking, and allowance is made for only one point of view. Questions are defined as legitimate or illegitimate within a rigid hierarchy, which molds the framework for discussion and debate in ways that make challenging injustice difficult. All of this poses a difficult conundrum, because it is the "exceptional" person, not the "ordinary" individual, who challenges the norms and standards in such contexts.

Instructing others to kill and maim at the SOA is a bureaucratic, highly regulated affair that takes place in classrooms and on training grounds, and U.S. military doctrine shapes all aspects of instruction at the SOA, as it legitimizes open aggression against certain kinds of people. Much like religious dogma, it refers to particular principles and policies about the way things are done. It dictates everything from how to initiate an ambush—with a single shot—to broader geopolitical visions that define who is, and who is not, the enemy. SOA instructors must adhere to doctrine as laid down in a variety of training manuals that are written at Fort Levenworth and translated into Spanish at the SOA, and those who extemporize too freely or stray from the prescribed orthodoxy are sanctioned and removed from the classroom. In the years since the discovery of torture manuals used at the School, which, Weidner claims, never represented U.S. doctrine, SOA officials are careful to screen training materials sent from Fort Levenworth and to remove anything deemed too sensitive to the politically charged context of the School. What remains are manuals that reflect a sanitized version of U.S. doctrine and that aim to produce "professional soldiers" who employ force legitimately in the context of "just wars." Teaching others to kill "by the book" is thus approved by the bureaucratic apparatus in which many participate but none bear responsibility.[7]

Finally, the well-being of ordinary men and women who bear the brunt of counterinsurgency warfare does not become part of painful moral dilemmas at the SOA. Considerable physical and social distance separates Latin American peasants, indigenous peoples, and urban dwellers from the lives of SOA officials and their trainees. The distance necessary to dehumanize powerless people, and remove them from any common ground with dominant groups, has been created to a considerable degree by the military itself. The naked violence unleashed against defenseless people during counterinsurgency campaigns, and the economic brutalization perpetrated by unpopular government policies that are upheld by the military, wreak havoc on peoples' lives by depriving them of the means to take care of themselves. Demeaning stereotypes of the "dirty," "lazy," "ignorant," and "violent" Other are thus constantly recreated. Poverty becomes defined as a natural condition—not the legacy of decades of enforced inequality—and it testifies to the slothfulness of the poor. Demands for better living conditions constitute evidence for the military of an innate propensity for violence and unruliness among peasants and indigenous and working-class people.

Such is the view of the indigenous Maya held by the Guatemalan military and the mestizo middle and upper class (see Schirmer 1998), and it is reflected in Colonel Weidner's understanding of postwar Guatemala and its unresolved problems. "There is," he says,

> a strain of incomprehensible violence in Guatemalan rural society that is a factor out there in [the process of] reconciliation that transcends the political issue. [Dealing with it] certainly goes beyond military competencies. I do not think that you deal with it successfully by weakening the legitimacy of the state to enforce the law. . . . I'm not a great expert on Mayan sociology but certainly you have history and culture going back to pre-Columbian times of violence as an accepted tool of society. The idea of human sacrifice and the indigenous concepts of war and taking prisoners. Do these things get erased by the Conquista and the colonial period and the nation-building period? We have seen a tremendous preservation of Mayan isolation from society. I would say that these things have not been eradicated.

The like-mindedness of Weidner and the Guatemalan military is in fact shocking, considering that this racist, anti-indigenous reasoning provided much of the basis for the Guatemalan military's genocidal campaign against indigenous people. "Mayan violence" is not only an inherent quality of indigenous people in this formulation. It is impermeable to social manipulation, having survived intact through centuries of colonialism and nation building. It even "goes beyond military competencies" to control, a fact that suggests a need for constant military readiness to fight outbreaks of indigenous savagery, and it has no connection to decades of brutalization by the Guatemalan military and its U.S. allies or economic exploitation by local elites.

A complex amalgam of loyalty, monopolistic authority, social distancing, and free-floating bureaucratic responsibility that is never assigned to specific individuals shapes the relationship of men like the commandant to the SOA. They pave the road between military training schools and the massacre sites of Latin America with bureaucratic activity and fill the signposts with clichés. Dismissal of Weidner's defense of the institution as simple dissembling is too simplistic, because it misses the way that particular ideologies are shaped in specific social and institutional contexts and become part of deeply held personal feelings. Weidner believed in his mission, and he never lost sight of his job. He also planned to retire from the military when his term as commandant ended. As his career wound down, challenging the military would have raised questions about his entire life, but, by continuing to do his duty, Weidner brought one chapter of his life to a close as he contemplated the start of another.

On one occasion, he invited me to a luncheon with other SOA students at the officers' club, where some fifty officers sat at large, round tables in the dining hall. The protocol officer seated me at the commandant's table, next to the colonel, and during the meal, while others guarded a respectful silence, Weidner described to me the difficulty of "getting the truth out about the School." He explained that, as an army officer, he could not lobby the government or accuse congressmen, such as Representative Moakley, of mendacity or stupidity, but he nevertheless played the lobbying game "pretty close to the line." He did so, he said, by inviting academics, students, human rights activists, and anyone else who was interested to come to the School. When I asked if he was lobbying me, he learned over and, in a mock whisper, said that if he encouraged me to call my congressman and urge him to vote in a particular way, that would be lobbying.

Weidner certainly knew that I was talking with members of SOA Watch, the human rights organization founded by Father Bourgois that spearheaded the movement to close the School, and he wanted to counter its view with his own. He knew, too, that torture was not taught at the School of the Americas at the time. After years of protests, several investigations, and a major controversy over the torture materials, the seamier aspects of military training had moved to less controversial military

schools and to Latin America. Weidner could rest assured that my presence in the classrooms of the SOA would not lead to a gruesome exposé of
grisly interrogation techniques.

Although considerable impression management always shaped our
interactions, Weidner and I established some common ground on which
to talk. We were both the middle-class products of white, Anglo-Saxon,
Protestant America, and we could—at least for a time—operate within
the traditional conventions assigned to our genders. Colonel Weidner
liked to talk and felt comfortable holding forth, and, like many women, I
knew how to listen. We both enjoyed discussing politics and Latin America, and he always seemed available for conversation. He was well read
and given to rambling philosophical exegeses on violence, democracy,
and the mission of the armed forces in a democratic society, and he liked
to quote the work of the right-wing, national defense intellectual Samuel
Huntington. Weidner was a former fellow at Harvard's Weatherford Center for International Affairs, and he once expressed to me a desire to enter
academia after his impending retirement from the army, but he feared
that his association with the SOA would lock him out of the academy. At
the time, I felt a twinge of guilty sympathy for him.

Research at the School of the Americas, where my military interlocu-
tors had a stake in making me adopt their interpretations, presents certain
pitfalls. One is what Antonius Robben calls "ethnographic seduction," a
concept he uses to refer to the disarming of ethnographers and their
dissuasion from critical inquiry through the subtle molding of appearances in interview contexts. Ethnographic seduction is, according to Robben, based on emotion, appearances, and the illusion of congeniality,
rather than reasoned argument, and it is easily confused with good rapport. He argues that this aspect of fieldwork, which may be unconscious,
is particularly problematic in research on political violence, where verify-
ing conflicting claims and assertions is impossible. The danger of ethnographic seduction is that one may "be led astray from an intended
course." One becomes so familiar with the power holders and the institutions of domination, or temporarily dissuaded by their appearances and
confused by their jargon, that the authority and destructiveness embodied in them is obscured from view (Robben 1995).

This was a risk that extended well beyond the interview context. SOA
officials made no effort to restrict me to Humanitarian De-Mining and

other courses that they wanted me to see. The commandant in fact reiterated to me on several occasions that I could sit in on any course that was in session, or accompany students to the firing ranges. The classroom-based courses were often mind-numbingly tedious and boring beyond any of my expectations. Frequently, I found myself glancing at my watch in eager anticipation of the ten-minute breaks that punctuated every hour of class time and offered brief respites, when it was possible to chat with the students around the coffeepot. After spending considerable time in the Command and General Staff Officer course, for example, I shared the opinion of one tired participant, who complained that it was just an endless repetition of dull slides with a series of bulleted points that were repeated with little variation by the instructor. "Slide, slide, slide, test. Slide, slide, slide, test. Slide, slide, slide, test." This was his description of how the days and the weeks passed by.

The classes were lessons in what Hannah Arendt has described as the "fearsome, word-and-thought-defying banality of evil" (1994, 252). They were deceptive in their capacity to shroud military training in a haze of tedium, triteness, and bureaucracy and to separate it from the people training at the institution, their relationships to each other, and the contexts in which they deployed their military skills. A more complete understanding required moving outside the classroom, examining the soa's history, and pursuing the connections that linked the institution, its officials, and students to people and events in Latin America, past and present.

CHAPTER 3

Foot Soldiers of the U.S. Empire

Reflecting on the School of the Americas, retired general and former U.S. Southern Command chief Paul Gorman opined that "an officer who has gone to the School of the Americas is 100 percent easier to do business with than somebody who has not, [because] graduates understand Americans. They understand that when Americans tell them that something is going to happen, we mean it." The crusty old general spent the better part of an illustrious career telling a variety of people, mostly from the Third World, that "something is going to happen," and sometimes it did. Gorman's long military career began in World War II, spanned the Korean and Vietnam wars, passed through the Pentagon and the CIA, and ended during the Central American civil wars of the 1980s, when he retired in 1985 as a four-star general. His years in the military coincided with the emergence of the United States as a global power and the projection of its power and technological capacity for destruction around the world. Not surprisingly, Gorman's arrogant words emerged from a U.S.-centered logic of empire, and they reflected the one-sided thinking of an officer who carried out his duties during a period of unprecedented U.S. global expansion.

The United States rarely acted alone in Latin America. Its dominance in the aftermath of World War II was based less on unilateralism— "tell[ing] them that something is going to happen"—than on a complex process of collusion in which U.S. military power was effectively joined with the indigenous military establishments of Latin America (Holden 1993). Particularly in the historically divided countries of Central America—the proverbial "backyard" of the United States—new, national-level military and police institutions created by the United States increasingly monopolized the use of large-scale violence by the mid-twentieth century, and states were more centralized than in the past. Factional warfare waged for decades by regional strongmen throughout the nineteenth and twentieth centuries had subsided. The centralization of coercive institutions displaced the exercise of violence from local bosses to national-level power holders in countries where war making and state making were closely intertwined. The consolidation of power and the increased capacity of states to commit violence grew out of multiple struggles that eliminated regional competitors and contenders within the state itself (Holden 1996; Tilly 1985; Stanley 1996).

The United States aided and abetted this process by arming preferred leaders, intervening militarily to prop them up when necessary, and using threats, loans, diplomatic pressure, and other techniques to control governments in power. Military strongmen and factions of the national elites developed relations of interest and patronage that tied them to U.S. centers of military and financial power. The result was the increasing internationalization of state-sponsored violence. National security apparatuses were better able to deploy lethal violence, and the development of dense personal networks between U.S. and Latin American militaries and police forces permitted the United States to exert greater control over its Latin American allies (Holden 1996; Huggins 1998). The School of the Americas aided this process by training national armies and knitting them together under the aegis of the United States.

The union of U.S. and Latin American military power did not happen the same way everywhere. National liberation struggles effectively disrupted it in some instances. The Cuban Revolution and the Sandinista takeover in Nicaragua seriously challenged U.S. power, while populist, and sometimes anti-imperialist, military regimes, such as those headed by Juan Velasco (1968–1975) in Peru and Omar Torrijos (1969–1981) in Panama, gave U.S. strategists headaches. Constant cleavages within national militaries, periodic disputes between Latin American states, and shifting social dynamics and changing power balances within particular countries also posed continuous threats to U.S. hegemony and the local status quo. All of this required continual vigilance. Although the SOA was never the only place that Latin Americans received U.S. military training, it was an important setting where, starting after World War II, the United States procured the cooperation of Latin American militaries in its imperialist project.

The School provides a useful starting point for exploring the "field of force" within which the militaries of the Americas constructed shifting transnational relations of power, inequality, and self-interest. Such an analysis must necessarily take place within a single analytic field. The understandings and relationships that connected outposts of hemispheric and local power shifted constantly within the highly contingent and differentiated context of global realpolitik, but through field-based instruction, classroom lectures, the subtle manipulation of cultural practices, and displays of technological power, the School brought together dispersed, and sometimes antagonistic, security forces and molded them into proxy armies under U.S. tutelage.

The cold war provided the justification for this process. While the United States dedicated itself to containing the global expansion of the Soviet Union and persecuted domestic critics with varying degrees of intensity, it assigned its Latin American allies the task of guarding against the threat posed by "internal subversion." National security doctrine (NSD) provided the broad rationale for fighting communists by assigning the maintenance of internal "order" to Latin American security forces, and by delegating to the United States the task of guarding the ramparts of the Western Hemisphere from external aggression. Concerns about "internal subversion," however, were not new to security forces and elites in Latin America, where variants of NSD developed in several countries. In Brazil, for example, authoritarian political leaders recognized the importance of internal policing for the stability of their regimes even before the advent of the cold war (Huggins 1998), and, in 1949, the military set up the *Escola Superior de Guera,* a training center that played a central part in the development of national security ideology. Guatemala and Argentina also crafted notions about national security that responded to perceived domestic threats and that did not follow U.S. thinking in lock-step fashion (Schirmer 1998; Andersen 1993).

Yet virtually everywhere, so-called subversives included peasants, workers, students, and others who increasingly demanded real agrarian reforms, better wages and working conditions, improved education and health care. The United States and local elites branded them "communists" during the cold war, and it was the "Communist Menace"—perceived as a threat to both the United States and Latin America—that propelled the military training of Latin American security forces by the United States. The United States directed enormous amounts of financial and material aid to Latin American militaries, and it transformed them into larger, better-trained, and more efficient national-level forces charged with eternal vigilance against what was perceived as an insidious, often invisible enemy that sought to wreak havoc throughout the body politic.

The cold war thus evolved into a permanent state of mobilization and vigilance against challenges to the status quo, and it spawned astronomical military budgets. Average annual U.S. spending during the forty-five years of the cold war (1945–91) in 2002 dollars totaled $351.8 billion (CDI 2002, 35), monies that were sucked from an array of possible social welfare initiatives. With the exception of those connected to the military, most U.S. citizens experienced the cold war as "peace" or witnessed it indirectly through televised images of the Vietnam War. Needless to say, this was not the case for the Vietnamese or many Latin Americans who endured its violence and destruction in chilling and gruesome ways. For these people, the cold war represented less a brush fire on the periphery of the industrial, capitalist world than a bloody orgy of violence in which thousands died at the hands of security forces trained by the United States.

Out of the Ashes of World War II

The Latin American Ground School—a precursor to the SOA—was founded in the Panama Canal Zone in 1946. It occupied a few buildings at Fort Amador in the Atlantic sector of the Panama Canal Zone, and ten officers and twenty-seven enlisted men ran its training program. The establishment of the Ground School coincided with renewed U.S. expansionist ambitions in the Americas and partially filled a power vacuum created by World War II, which ruptured long-standing military ties between European imperial powers—particularly France, Italy, and Germany—and Latin America. The war obliged the European powers to concentrate on waging war at home, and it forced them to close their Latin American military missions, which had equipped, trained, and organized armies almost everywhere except Panama and Nicaragua. At war's end, the Europeans confronted the task of reconstruction, which took precedence over the renewal of ties with Latin America, and the victorious and relatively unscathed United States moved to fill the void left

Figure 3. The fascist salute—enduring symbol of German military training, 1991. Photo by Jeremy Bigwood.

by the Europeans and to consolidate its position as a global superpower. Through its own military missions, the United States started to arm and train Latin American security forces,[1] but the initiative met a mixed reception from many Latin American commanders. A number of military leaders had long looked to Europe for advisors and training: the Argentines, for example, had ties to the German military that dated back to the Bismarck era. European training was perceived as superior to the instruction offered by the Americans in the Panama Canal Zone.

During the late 1940s and 1950s, the stated goal of U.S. military strategy in Latin America was to train and equip security forces to repel an attack by a nonhemispheric power, particularly the Soviet Union. Together with twenty Latin American countries, the United States signed the Inter-American Treaty of Reciprocal Assistance, better known as the Rio Treaty, in 1947. The Rio Treaty was one of several mutual defense agreements formulated by the United States, and its main purpose was to organize a united front against an attack (Schoultz 1987, 179–80). Yet as Holden indicates, "military grant aid was extended to Latin America not to defend it from attack but to help encourage a cooperative attitude" (1993, 29). The real agenda was less "hemispheric defense" than a far more traditional concern: the maintenance of U.S. dominance in the Americas and the control over raw materials. This project required political stability and governments willing to cooperate with the United States, and, perhaps more importantly, it entailed suppressing the domestic opponents of friendly regimes (Holden 1993). The provision of arms became an important means through which the United States purchased the collaboration of militaries, enhanced their power vis-à-vis domestic challengers, and kept nonhemispheric contenders and weapons suppliers at bay. The commanding general of the Caribbean Defense Command understood the importance of arms sales very well when he noted in a 1945 memo to the War Department that

> it is an established fact that if equipment to supply these armies is not forthcoming in the near future, we face the probability that Mission contacts will not be renewed. It took the war and the elimination of our chief competitors to put us in the unique position we occupy today. Immediate action must be taken to furnish arms and equipment to our Latin American neighbors. Even now we are confronted with the probability of losing the advantage we have gained at high cost and much effort.[2]

The general did not have to worry. Although Latin American commanders had complained of difficulties in securing armaments during World War II, the postwar era created more favorable conditions for arms sales to the region. Defense manufacturers sought out new markets for their wares, and Congress created generous military aid programs. Facilitating arms transfers not only helped to secure U.S. access to raw materials and the general cooperation of regional militaries, it also tied Latin American militaries to the use, and thus the continued purchase, of technology produced in the United States. The defense establishment referred to the latter as "standardizing" Latin American militaries, and the Caribbean Defense commander put it bluntly when he stated that "standardizing" Latin American armies with U.S. equipment furthered the "penetration of the United States into the military system of any country so that such nation becomes dependent on us."[3]

Yet arming Latin American militaries was only part of the process of buying their cooperation and insuring their subordination to the United States. Soldiers had to learn how to use the equipment, and they needed to understand and identify with the imperial goals of the United States.

The Latin American Ground School addressed these needs. Its activities supplemented the work of the U.S. Army missions in Central and South America and centralized to a considerable degree the military instruction of Latin Americans, who, since at least 1939, had trained at a variety of military bases in the Panama Canal Zone.[4] The Ground School, however, did much more than train students in the tactics of warfare. It initiated their incorporation into the ideology of the "American way of life" by steeping them in a vision of empire that identified their aspirations with those of the United States, a process that, as we saw in chapter 1, continues today. A special training program for Argentine officers and enlisted men demonstrated this dual process of technical training and cultural persuasion.

Fifteen army officers and twenty enlisted arrived at Fort Amador in May 1948. They came for a three-month course on the operation and maintenance of fifty 90-millimeter antiaircraft guns that Argentina had just purchased from the United States. Unlike groups of students who arrived in later years, the Argentines encountered a school that was struggling for survival and eager to make a name for itself. Classes were constantly underenrolled because the United States, embroiled in the Korean War and concerned with rebuilding war-torn Europe, did not yet provide the financial support to Latin American militaries that it would in subsequent years. Many Latin American commanders preferred to use limited funds to send their rising stars to military service schools in the "Zone of the Interior," a term used among U.S. global strategists to refer to the continental United States. They believed that the instruction was superior at these service schools because their men received the same training as U.S. servicemen. Consequently, the Ground School struggled under a shadow of doubt to establish itself as a viable training center in the 1940s and 1950s, and its future was never certain.

U.S. Army officials therefore felt a good deal of pressure to make the Argentine program a success and to demonstrate to critics within the U.S. military establishment that the Ground School was, indeed, exposing trainees to the American way of life. Their training mission also formed part of a broader U.S. initiative to "capture" the Argentine military, which, under Juan Perón, had leaned toward the Axis during World War II.[5] "We are all into this [program] up to our necks," fretted one officer during a special staff meeting on the Argentine students. "There is no division of responsibility. The whole thing is on the shoulders of Col. Nelson but we must give him all the assistance we can from all angles."[6] The U.S. officers and trainers awaited the Argentines with nervous anticipation. They dwelled on every aspect of the training program, which, they believed, carried considerable symbolic weight for the United States. One individual effused that "it is no exaggeration to state that the cooperation and the solidarity of the Western Hemisphere nations depends to a great degree upon the impressions which the Argentine personnel take back with them to their native country."[7] Army officers described the Argentines as "extremely high type personnel [who] are probably well qualified."[8] They told everyone who interacted with them to learn the customs of the Argentine army and to recognize the insignia used to designate particular ranks. Most importantly, the officials advised course instructors from the Sixty-Fifth Antiaircraft Artillery Group to instill among the trainees faith in the weapons, and, in this way, to draw a connection between the powerful weapons and the United States. They asserted that, if the Argentines "had confidence in the weapons, they would have confidence in the United States," and this confidence would spread when the Argentines returned home and taught others how to use the guns.[9]

Building confidence in the United States was a complex undertaking that involved much more than impressive weaponry. The Argentine trainees held strong opinions about their national and racial superiority vis-à-vis other Latin Americans, and these views did not go unappreciated by their North American hosts. U.S. officials manipulated race and the politics of culture to sell the Argentines on the power and superiority of the United States and to make them feel equal to their imperial overlords. They deemed it essential that "in view of the nationalistic feelings of the Argentine[s] . . . and their belief in certain racial theories, . . . they be made to feel that they enjoy equal privileges with American officers and enlisted men."[10] They issued the Argentines passes that distinguished them from other students at the Latin American Ground School and that granted them exceptional privileges, such as special commissary rights and exemption from the maintenance duties carried out by darker-skinned Latin Americans.

The U.S. military in Panama had its own racial theories that drew on the experiences of nineteenth-century British and French colonizers in tropical climates. Whites, it was believed, suffered more from prolonged exposure to intense heat and humidity than dark-skinned peoples. They were also allegedly more susceptible to certain diseases, such as yellow fever. Some physicians even asserted that the tropics stunted the growth

Figure 4. A 1950s Army cartoon depicts white fears of racial degeneration in the Panamanian tropics. Z.I. refers to the "Zone of the Interior," a military term for the continental United States. From U.S. Army Caribbean Status Report, Nov. 1951. Courtesy of National Archives, Washington, D.C.

of white children, who faced additional risk of degeneration from too much association with "the natives." Whites therefore required periodic vacations and reassignments to cooler northern climates in order to restore their racial and physical vigor, and not surprisingly, dark-skinned peoples assumed the burden of hard manual labor in the Canal Zone.[11]

The army provided softball and soccer equipment to the Argentine trainees so that they got exercise and developed relationships with members of the Sixty-Fifth Antiaircraft Artillery Group, who worked and lived with them. The Argentines even received golf clubs, because, as one official lamented, "It would be tragic if the students were Golfers and did not bring their equipment with them and then found themselves living and working on a golf course."[12] Such a sensibility probably reflected less the blatant ethnocentrism of the official than evidence that he had done his homework. Anglophilia ran deep among certain sectors of the Argentine armed forces, and it was likely that some officers did, indeed, play golf in 1940s Argentina. Whether they actually found time to practice their golf swing in Panama, however, was unclear.

School officials made every effort to keep the students occupied at all times. They did so for two reasons: first, they hoped to convey to the students a particular vision of U.S. citizens as industrious and successful; second, they wanted to keep students in the Canal Zone and out of

Panama. Officials worried that disorders created by students, such as public drunkenness, expressions of immorality, or fights, would provoke the ire of Panamanian authorities and cause public relations difficulties for the U.S. military. The commandant did not mince words when he told a group of staff officers that

> In addition to taking care of these people and making them welcome and happy while here, they must be kept busy. Organize their instructions—make the schedule so that they . . . do not have too much free time. Give them organized athletics so that they will stay in the Zone and out of Panama. They are here to learn 90mm AAA [antiaircraft artillery] equipment and technique. They must carry with them the impression that this is the way we work and that is why we are a great nation.[13]

Although the Ground School tried to portray U.S. citizens as hard working, upstanding, and industrious, its efforts to keep students happy often sent a mixed message about Yankee morality. This is evident from the ambiguous signals students received from the U.S. military about prostitution. On the one hand, school officials prohibited all personnel—U.S. and Latin American—from entering any establishment in Panama defined as a house of prostitution, and they cited the army's 1946 Repression of Prostitution Act to emphasize their concern for the morals, health, and welfare of all service personnel. On the other hand, they distributed an information booklet to Argentine officers that listed—on seven, single-spaced, typed pages—the names and street addresses of every brothel and strip joint in Panama City, Colón, and areas described as "outlying districts." In Colón, for example, the army prohibited visits to addresses No. 12184, No. 11185, No. 11190, and No. 2019 on 12th Street. Similarly, on E. 17th Street in Panama City, "Matilde's Place and house of prostitution under her place at No. 10" were off limits. The booklet—intended for a much broader military audience than just the Argentines—then went on to provide a listing of similar establishments in the capital cities of Nicaragua, Guatemala, and Costa Rica.[14] Its specificity reflected not only the army's obsession with detail but also an intimate knowledge of the seamier side of local life drawn perhaps from direct experience. To head off any potential embarrassment for the Argentines and themselves, U.S. officials notified the Canal Zone police of the Argentine trainees' "exceptional status," and they made overtures to the Panamanian police to obtain their cooperation in "affording the Argentine personnel every courtesy and consideration."[15]

It is not clear how the Argentine trainees evaluated their experience at the Latin American Ground School, or how they viewed their North American instructors and hosts. Clearly, U.S. officials were eager to accommodate the Argentines and anxious that these students and others like them take home a view of the United States as enterprising, efficient, and powerful. They did so by pandering to racist beliefs, extending special privileges to the Argentines that other Latin American trainees did not enjoy, and demonstrating a mastery of powerful weaponry that, they hoped, communicated a strong message about the awesome strength of the United States and the no-nonsense, resolute nature of its people. The practices represented some of the ways in which the United States identified the aspirations, desires, and conceits of carefully chosen subordinates and tied them to its own vision of imperial rule.

Integrating trainees into an imperial military required that Ground School instructors and bureaucrats attend constantly to the politics of race and rank. They were particularly challenged in the school's early days by the practice—born of necessity—of training officers and enlisted men together. Because of financial constraints and low course enrollments, officers and enlisted men, who reflected race and class divisions within the military, frequently found themselves in the same classes; a practice that constituted "fraternization" and posed a threat to military discipline at the Ground School. The commandant, like most of his colleagues, felt that the practice also eroded the quality of education. In 1947, then commandant E. M. Benitez wrote to his superior about the issue. "In the weapons course," he noted, "there is a major from Guatemala with fifteen years of military service and a private from Honduras with four months service." To complicate matters, the ten students in the course averaged nine years of formal education, but they ranged from the Honduran private, who had only completed the fifth grade, to a Venezuelan lieutenant, who had graduated from high school. The commandant urged that, "for both training and disciplinary problems presented by the Latin American students," separate courses for officers and enlisted men be instated.

The commander-in-chief of the Caribbean command took his recommendation under consideration and solicited the views of all the U.S. military mission chiefs posted around Latin America. The chiefs were unanimous in their support for separate courses, and, in several cases, based their conclusions on strong racial and class preferences that emerged from consultations with local commanders and their own experiences in the U.S. Army, where intelligence testing was used as early as World War I to justify racial hierarchies.[16] Colonel Harry W. Miller, chief of the U.S. military mission in Paraguay, supported the separation of officers and enlisted men because he believed that "there is a great mental difference between these groups." In Paraguay, those who did not measure up mentally were, of course, the largely indigenous troops who composed the lowest ranks of the army. Lieutenant Colonel James Adams, the head of the Honduran mission, felt that "no course at the Latin American Ground School can be equally beneficial to the most advanced and most retarded students." And mission heads in Bolivia and Guatemala, where indigenous people filled the lowest levels, noted the "caste system" in these countries and used it as a justification for keeping officers and enlisted men separate. The mission commander in Ecuador referred to a "class division" and commented that it was even greater in the Ecuadorean army than in the U.S. armed forces.[17] Given these opinions, it should come as no surprise that officers and enlisted men eventually found themselves receiving separate classes.

The Ground School—like the U.S. Army in general—also advanced different visions of empire to officers and enlisted men. Officers, for example, imbibed a paternalistic ideology of leadership that exuded the middlebrow earnestness and social-climbing ambition of white, middle-class managers. U.S. Army publications emanating from the United States heartland at Fort Levenworth spelled out this ideology. According to one of these manuals, leadership involved "influencing and directing others to an assigned goal in such a way as to obtain their obedience, confidence, respect and loyal cooperation," and it required a number of fundamental personal attributes that included knowing yourself and seeking self-improvement, taking responsibility for your actions, developing responsibility in others, and setting an example. A good leader was therefore a self-controlled striver who made thoughtful decisions and managed others so that they willingly and loyally accomplished any task assigned to them.[18]

U.S. officials constantly evaluated Latin American trainees on a range of personal attributes linked ostensibly to leadership capacity that had little or nothing to do with the courses that students took. Major John Stengler, the Director of Training, rated the Bolivian second lieutenant Bernardo Anglarill for his performance in a five-week basic engineering course on the basis of fifteen criteria that included tact, bearing, dependability, enthusiasm, resourcefulness, sense of responsibility, judgment, industry, initiative, and cooperation. Although Anglarill received an overall rating of satisfactory, the major described him as "frequently lazy, offering no inspiration to others, somewhat antagonistic and ignores suggestions."[19]

If subordinates like Anglarill remained unpersuaded through appeals to their reason, then the "judicious use of punishment" became an important way to instill proper behavior. Good leadership rested on discipline, which, for the military, constituted the extension into its ranks of the order and control that characterized any well-organized society. A disciplined unit—like a properly disciplined society—thus displayed "the cheerful subordination of the will of the individual for the good of the group."[20] Some "individuals," however, were by nature more subordinate than others.

Most of the trainees—officers and enlisted—who passed through the institution wrote glowing reports to U.S. military mission commanders in their home countries. They expressed considerable delight with the opportunity to use modern, U.S.-made weapons and to bask in the refracted glow of U.S. power. Some recruits were barely literate, and for many young men, training at the Latin American Ground School represented an opportunity to hobnob with one of the most powerful militaries in the world, and to travel for the first time to another country. The importance of this experience was not lost on the mission chiefs. Lieutenant Colonel Edwin Messinger, who headed the U.S. mission in Costa Rica—a country without an army—commented that "although the things that are learned at the [Ground School] are considered to be of special value and assistance to this Mission in its training program, it is believed by far that the greatest contribution that [the Ground School] makes is that after attending, the students from Costa Rica come back and express to their countrymen the admiration and enthusiasm they feel for the United States. Particularly, they are impressed with the efficiency and the equipment of the United States Army. . . . the Military Mission believes this good-will will pay far more dividends to the United States than any amount of military training."[21]

Yet despite the trainees' favorable reviews, tensions at the Latin American Ground School between those who increasingly wielded power in the hemisphere and those who could only aspire to a small stake in it were never far from the surface. Salvadoran trainees, for example, complained of the "superior attitude" of their U.S. instructors in a basic weapons course,[22] and, on another occasion, Peruvian students criticized U.S. trainers who consistently discouraged questions and appeared unconcerned about their progress. The Peruvians were particularly incensed when these instructors staged a question period to impress a group of visiting Latin American officers. The instructors convened the class on a holiday, received student questions while the visitors passed through, and then quickly dismissed everyone.[23]

Some 7,886 Latin American students trained at the school between its founding in 1946 and the eve of the 1959 Cuban revolution. Almost half (42 percent) of these trainees came from the small Central American countries of Honduras, El Salvador, Guatemala, Panama, and Nicaragua, but nearly a quarter of all students (1,896) came from Nicaragua, where the Somoza family dictatorship rose to power in 1936.[24] The Nicaraguans were concentrated in three broad course areas: communications, weapons and tactics, and military police training. Communications courses focused on the operation, maintenance, and repair of radios; weapons instruction dealt with the use of both small arms and heavy artillery; and the military police course contributed to the modernization of Somoza's national guard, created by U.S. Marines in the 1930s to repress the nationalist insurgency of Augusto César Sandino.

When they finished training, the Nicaraguans returned to a country torn by massive social dislocations set in motion by the expansion of cotton cultivation. Landed oligarchs, urban professionals, Somocista politicians, and national guard officers, intent on joining the agrarian bourgeoisie, were investing in cotton cultivation and cashing in on high world prices. Cotton fever fueled a demand for land, and peasants, who had a variety of tenancy arrangements with landlords, found their access to hacienda property dramatically limited. They also experienced an erosion in the quality and stability of agricultural labor. Not surprisingly, social discontentment seethed in rual Nicaragua, and the job of containing and repressing it fell to the national guard, which increasingly acquired a reputation for brutality (Gould 1990).

In addition to the guard's outsized presence at the Ground School in the 1950s, the Nicaraguan military, and the national guard in particular, benefitted from more U.S. military aid grants and arms sales than any other Latin American state. The aid not only bolstered national guard efforts to control the internal, anti-Somocista opposition but also sustained Nicaragua during a period of intense hostilities with neighboring Costa Rica. Costa Rica did not have an army, and even though its president, José Figueres, was a freely elected social democrat, the United States preferred the Somoza family dictatorship to democratic Costa Rica, which received comparatively little U.S. assistance (Holden 1993).

The 1959 Cuban revolution did not change the U.S. preference for arming dictators, but it permanently altered the political landscape of Latin America. The revolutionary movement led by Fidel Castro and his ragtag band of guerrillas toppled the repressive Batista regime and then consolidated its own power base by radically reconfiguring Cuban society. The revolutionaries expropriated the resources of wealthy Cuban capitalists and their U.S. allies. They launched a radical agrarian reform program, and they initiated far-reaching changes in health care and education. Then, the government embraced Marxism-Leninism and allied with the Soviet Union. The United States perceived the Cuban revolution as not only a threat to its interests in Cuba but also to its hegemony throughout the hemisphere, where the revolution represented an appealing alternative to the status quo for many Latin Americans. The revolutionaries themselves encouraged these perceptions by announcing their intention to spread revolution across Latin America, and, throughout the region in the 1960s and early 1970s, small groups in almost every country embraced the Cuban example and tried to emulate its tactics and strategy (Pérez 1995; Castañeda 1993, 68–69).

The U.S. military believed that the Cuban revolution provided clear proof of an international communist conspiracy, and national security doctrine formed the bedrock of military thinking in its aftermath. The subsequent inability of the United States to impose order on Vietnam, despite the commitment of over 500,000 troops, almost certainly played a part in convincing military strategists of the ungovernability of the Third World, and it probably magnified fears of conspiring communists in Latin American peasant villages. Communist subversion became defined as anything that challenged the status quo, and broad sectors of the population—students, activists, trade unionists, peasant organizers, and religious catechists—came under suspicion.

The triumph of the Cuban revolution insured the institutional survival of the Ground School. After undergoing a reorganization in 1949 and reemerging as the U.S. Caribbean School, the institution was restructured again in the aftermath of the Cuban revolution and became known as the

School of the Americas in 1963. Training the militaries of the Americas to fight communism became its raison d'etre, and counterinsurgency instruction evolved as a major new focus of its activities. The School had moved into a former hospital building on the grounds of Fort Gulick on the Atlantic side of the Panama Canal Zone, and 13,500 students attended its classes in the decade following the Cuban revolution, a 42 percent increase over the first thirteen years of the School's existence. Venezuela, Nicaragua, Bolivia, Panama, and Peru were the countries most heavily represented. Each sent between one thousand and two thousand trainees to the school between 1960 and 1969.

Cecil Himes was the commandant of the School when the Cuban revolutionaries stormed into Havana, and he oversaw the organization of the first counterinsurgency training course. "President Kennedy issued instructions that we would have a counter-guerrilla course," he told me in 1999, when I interviewed the eighty-three-year-old retired colonel at his home in a leafy suburb of Birmingham, Alabama. "The army had the Special Forces and [setting up the counter-insurgency program] was their mission. The Special Forces came down and made recommendations and assisted us in setting up the course. The [School] instructors also worked with the Special Forces people."

The military doctrine of counterinsurgency warfare started to dominate U.S. policy in Latin America after the Cuban revolution. Its special contribution, according to Michael McClintock, "was the legitimation of state terrorism as a means to confront dissent, subversion, and insurgency. The characteristic organizational forms of the counterinsurgency state were vast formations of paramilitary irregulars, elite Special Forces–style units, and powerful centralized intelligence agencies under military control" (1991, 121). Counterinsurgency warfare was the covert side of the Alliance for Progress, the massive U.S. assistance program to Latin America initiated by President Kennedy that combined highly publicized civic action programs with clandestine terror and massive violence. Civic action and security in fact went hand in hand. Many policymakers, academics, and military strategists in both the United States and Latin America believed that the military was the only institution capable of maintaining domestic tranquility while simultaneously promoting economic growth. Nevertheless, counterinsurgency doctrine's emphasis on development and security provided the armed forces with a rationale for intruding more deeply into the lives of ordinary people, and its most striking feature was that it prescribed terror as a tactic for fighting guer-rilla insurgencies, which, at least in the eyes of the U.S. military, used the same tactic against the governments of the region (McClintock 1991).

Counterinsurgency warfare fascinated President Kennedy, whose ad-ministration underwrote the resurgence and expansion of the Special Forces. Operating out of Fort Bragg, North Carolina, the Special Forces provided counterinsurgency training to foreign officers and those from the United States' four service branches, but most of its training of foreign forces was conducted outside the United States. The Mobile Training Team, or MTT, was—and continues to be—its principal medium for train-ing soldiers abroad. An MTT consisted of two officers and ten enlisted men sent on short-term missions to work with conventional armies, intel-ligence groups, and paramilitary irregulars. More than six hundred Spe-cial Forces MTTs operated in Latin America between 1962 and 1967 (Mc-Clintock 1992, 187).

A battalion of the Seventh Special Forces Group visited Fort Gulick in 1961, shortly before the first counterinsurgency course began at the school on July 31. At the same time, according to Himes, a group of anti-Castro mercenaries were training in Panama to overthrow the Cuban govern-ment and almost certainly receiving instruction from the Special Forces. The mercenaries were also helping themselves to the School's supplies. "I was aware that we were training forces to go on the [Bay of Pigs] inva-sion," he told me. "They were being trained in Panama but not with us [at the School]. They were drawing supplies from my people. We were or-dered to issue them things. All that I knew was that trucks were coming and getting supplies that were for us. They were Cubans, and they were training on the east side of the isthmus, [where] they evidently had some encampment of their own."

During Himes's tenure as commandant, he maintained cordial rela-tions with then-commander of the Nicaraguan National Guard, Anastasio Somoza Debayle, who expressed keen interest in counterinsurgency training. "General Somoza was very interested in the course," he told me. "They had trouble in Nicaragua, and he sent quite a few students down to us." Three Nicaraguan students attended the first course, which began July 31, 1961, and ended nine weeks later on October 6. They accompanied thirty-seven other Latin American students, who ranged in rank from second lieutenants to majors and represented eleven other countries. Over the next seventeen years, over one hundred Nicaraguan officers took the counterinsurgency course, which was variously labeled "Counter-Resistance" and "Irregular Warfare." Many more learned "irregular war-fare" skills in classes on small-group patrolling techniques, interrogation, psychological operations, and jungle warfare. When the Sandinista National Liberation Front overthrew the Somoza dictatorship in the bloody 1979 revolution, many of these men stood accused of gruesome human rights violations that included the torture, murder, and disappearance of thousands of people.

Nicaraguan students had figured prominently at the School through-out the 1960s and 1970s, when Somoza sent cadets from the Nicaraguan military academy to complete their final year of officer training at the SOA. Between 1960 and 1968, 218 Nicaraguan cadets, or an average of 24 stu-dents per year, took a 46-week course on basic military skills and tactics. As Colonel Himes explained, "Somoza was out of the class of '46 at West Point, and he set up a little West Point, or a military academy, in Nic-aragua. So the students had three years at their academy and their last year was with us. At the end of the year, they graduated and went back to their country as second lieutenants in their national guard." Nicaragua was just one of many other Latin American countries that emptied their military academies of senior cadets and shipped them to the School of the Amer-icas for polishing.

The cordial relationships that developed between men like Colonel Himes and Anastasio Somoza further greased the machinery of empire. Somoza was a frequent visitor at the School during Himes's three-year assignment as commandant, 1958–1961, and he even stayed at the School on occasion. Himes recalled that "sometimes I would come to my office in the morning, and my secretary would come in and say that General Somoza is having breakfast. So, I would go in and sit with him and have a cup of coffee. He'd come in during the night and found a room with an empty bed. Sometimes the way that I knew he was in Panama was be-cause my wife would receive a bouquet of flowers from him. I'd say, oh, the general's in Panama now." Somoza's visits to Panama also entailed more than checking up on his students and currying favor with men like Colonel Himes. Himes recalled that on one of his visits, Somoza men-tioned that he was going to the city of Colón "to roust this gal who was very communist . . . apparently a leader." He planned "to give her a hard time," according to Himes.

Himes, a U.S. colonel, and Somoza, a Nicaraguan general, negotiated a relationship within a rigid military order in which nationality and the relative power of one's institution could blur the distinctions of rank and the behavioral imperatives defined by them. Yet Somoza could, in many ways, operate more effectively in a transnational imperial context than Himes, whose command of Spanish was rudimentary, and whose knowledge of Nicaragua and Latin America was limited. Somoza spoke impeccable English, acquired from years spent in the United States, and he was married to a U.S. citizen. Both men were West Point alumni, and it was this distinction that, for Himes, represented an important connection between them. "We had a good relationship," Himes asserted, "because, after all, we were [West Point] academy graduates, and there is a certain bond there."

When Somoza organized a celebration to commemorate the twentieth anniversary of the Nicaraguan military academy, he invited Himes and his wife. The Himeses flew to Managua on an air force plane with another U.S. military couple on Somoza's guest list. Colonel Himes recalled the pomp and ceremony that surrounded the occasion, and he was particularly impressed by the fact that members of his small U.S. contingent did not need their passports to enter Nicaragua. "The army met us at the airport," he explained, "and escorted us to our quarters. Then, later, they took us back to the airport, so we didn't use the passport at all." One evening, during the three days that they were in Managua, the Himeses attended a formal military dance with high-ranking members of the Nicaraguan army and government, and the colonel got the opportunity to dance with Somoza's wife. "I danced with Mrs. Somoza," he recalled fondly. "I didn't realize that she was American, but she retained her citizenship. Imagine, dancing with the president's wife."[25]

Colonel Himes exulted in the pageantry of power. A native of Batavia, Ohio, where, he likes to say, "Cincinnati is the largest suburb," Himes reveled in the deference accorded him by others in Panama. "I was a big fish in a small pond," he told me. "As the commandant, I was the senior officer on the Atlantic side, the post commander, and the commandant— that's three titles. I had to go to all the diplomatic affairs. Everybody knew me, and I was greeted everywhere . . . when I left, [the city of] Colón made me an honorary citizen." Himes retired from the military in 1967 and settled in Birmingham, Alabama, where he ended his military career as the deputy core commander of the national reserve. He remained in Birmingham because he had many friends, acquired a house with a 5.25

percent mortgage, and could play golf all year. Yet despite its obvious pleasures and comforts, life in the Birmingham suburbs could not compare with the pomp and ostentation that filled Himes with feelings of self-importance in Panama, and it should therefore come as no surprise that he fondly recalled his time at the School as the best posting of his career.

Aiding Dictators and Suppressing Revolution

As the cold war intensified in the Americas after the Cuban revolution, different countries and their armed forces fell into and out of favor, and the students at the School and the courses they took reflected these changing dynamics. Cuba's presence ended abruptly after the overthrow of Batista, and Nicaragua's participation terminated twenty years later with Somoza's ouster by the Sandinistas. Nevertheless, U.S.-backed military dictatorships ruled most of Latin America throughout the 1970s, and their security forces were well represented at the SOA. Flush with U.S. military assistance, the heads of these armed forces found themselves in a position to send large numbers of troops to the School, and the creation, in 1976, of a mechanism for subsidizing the training of foreign soldiers, called the International Military Education and Training (IMET) program, facilitated the flow of soldiers to the SOA. Implemented by the Defense Department and funded through the foreign aid appropriation process, IMET paid for the training of international troops by providing grants to their governments. The respective armed forces then selected courses at U.S. military service schools for their personnel to attend (LAWG 1998), and governments footed the bill for travel and living expenses.

Soldiers from Bolivia, Chile, Colombia, Honduras, Panama, and Peru figured most prominently at the School during the 1970s. Between 1970 and 1979, each country sent between eleven hundred and eighteen hundred students to the SOA, and together they accounted for 63 percent of total enrollment. Bolivian General Hugo Banzer, who took power through a violent coup in 1971 and ruled until his downfall in 1978, was a SOA graduate. His penchant for brutality and his antidemocratic inclinations were probably not acquired when, as a young captain, he took a short course in 1956 to prepare him for duty as a driver. Banzer was, however, a long-time friend of the United States and so impressed the U.S. Army in his later career that it inducted him into the School's Hall of Fame in 1988.

Bolivian students flooded into the School of the Americas when Che Guevara launched an ill-fated guerrilla campaign in a remote southeastern region of the country in 1967. The guerrilla insurgency came at a time when army generals had their hands full with a militant tin miners' union and an organized peasantry. To deal with these and other potential threats, successive military regimes dispatched an average of 155 soldiers per year to the SOA between 1967 and 1979, an increase of 40 percent over the 1960–1966 period. The majority of the soldiers were senior cadets from Bolivia's military academy who spent nearly a year in the Panama Canal Zone studying the theory of communism, ground mobility tactics, intelligence, marksmanship, and how to use special purpose weapons and surveillance devices. Large numbers of Bolivian trainees also enrolled in a three-week class on jungle operations in which they studied the theory of guerrilla warfare and learned how to navigate and survive in the jungle.

More Chilean soldiers trained at the SOA between 1970 and 1975 than from any other country during the entire decade. This period coincided with the rise and fall of Salvador Allende, Latin America's first freely elected socialist president. Allende's Popular Unity government assumed power in 1970, but its experiment in socialist democracy ended in a bloody military coup d'etat on September 11, 1973. The United States aided and abetted the coup plotters and engaged in "massive covert interference" in Chilean affairs to destabilize the Allende government (Dinges and Landau 1980). It funneled money to the anti-Allende media, created a secret task force under Secretary of State Henry Kissinger to sabotage the government, and made every effort to woo the military to its way of thinking.

The presence of the Chilean military at the School of the Americas had increased over the 1950s and 1960s, and by the 1970s, the impact of so many years of training on the Chilean officer corps was readily apparent. Writing in his diary in November 1973, two months after the coup, General Carlos Pratts, an Allende loyalist who was killed by the Chilean secret police in a 1974 car bomb explosion in Buenos Aires, described how Chilean security forces confused the Chilean national interest with the interests of the United States:

> As far as the internal enemy is concerned the opinion acquired by those who have attended courses at the School of the Americas and others organized by the Pentagon has been increasingly prevalent. . . . Many of these [soldiers] have responded to the stereotypes and

thoughts which were inculcated into them during these courses and, believing they were liberating the country from the "internal enemy," have committed a crime which can only be explained by their ingenuousness, their ignorance and their political short-sightedness. . . . I used to tell the president that we should send our officers to know what it was like in the countries of Europe, Africa and Asia, not so as to copy or imitate their armed forces but so that they should widen their horizons and understand that the world does not begin or end in the schools of the Pentagon (cited in O'Shaughnessy 2000, 27–28).

One thousand five hundred and sixty Chilean soldiers attended the SOA between 1970 and 1975, but the majority (58 percent) came in the two years following the coup, when the military ruled and repression in Chile was most intense. Most of the trainees were low-ranking officers who were taught how to plan, execute, and control company-level operations, which included offensive, defensive, and psychological tactics. Another large group of approximately two hundred noncommissioned officers received instruction on how to lead and train small units of soldiers.

The fall of Salvador Allende and the establishment of the Pinochet dictatorship had implications for Peru and its military regime, which took power in 1968. Fearing aggression from a historically hostile neighbor, the Peruvian government increased annual defense expenditures from 7.2 percent between 1970 and 1974 to 22 percent between 1974 and 1977 (Cotler 1983, 27–28). Perhaps more importantly, however, military rulers had to contend with intensified popular mobilization that eluded their ability to control. Land invasions increased, as peasants struggled to bring about a just agrarian reform. Miners and teachers pushed their economic demands on the state, a new Cuban-inspired left emerged, and popular sectors increased their autonomy and organizational capacity (Cotler 1983).

Not surprisingly, Peruvian participation at the SOA peaked in the 1970s, despite the warm relations established between the reformist regime of General Juan Velasco (1968–75) and the Soviet Union. Of the 1,820 Peruvians trained at the SOA between 1970 and 1975, 62 percent were sent by the Velasco regime, while 38 percent trained during the more conservative period of U.S.-friendly General Francisco Morales Bermúdez (1975–80), who ousted Velasco. Nearly half of the trainees were concentrated in three overlapping courses: jungle operations, internal security operations, and basic combat and counterinsurgency, and over a third were officers dispatched to an orientation course for company-level commanders.

Further north on the Central American isthmus, Panama, under the leadership of General Omar Torrijos, dispatched 1,422 trainees over the course of the decade. Most of the soldiers were concentrated in four classes: jungle operations, leadership training for noncommissioned officers, irregular warfare, and intelligence. Following Torrijos's death in 1981, his former intelligence chief, General Manuel Noriega, seized power. Noriega had been on the U.S. Army's payroll since the late 1950s when he was a young cadet in the Peruvian military academy,[26] and he was also a proud graduate of the School of the Americas. When he was a first lieutenant, Noriega spent most of 1967 taking courses at the SOA. He participated in a basic course for infantry officers, studied combat intelligence and counterintelligence, and enrolled in the jungle operations class, which he had taken previously in 1965.

Retired U.S. General Fred Woerner—a former commander-in-chief of the U.S. Southern Command (1987–89)—worked with, around, and eventually against Noriega. He recalled that Noriega routinely wore the SOA crest on his Panamanian uniform. "That may not sound like much to you as an anthropologist," Woerner told me, "but it's as significant as an aboriginal tribe wearing a certain type of clothing or necklace. What you wear on your uniform is very important in a military context. You are a peacock strutting. Anyone else in the military can read a man's career on his uniform, and what he is saying is: I'm proud of these accomplishments. And there [the School crest] was, right [on Noriega's chest]."

Before his capture by invading U.S. soldiers, Noriega and the Panamanian military enjoyed a unique relationship with the SOA because of its location in Panama. If the School had a cancellation or could not fill all its courses, the United States regularly provided the seats in the course to the Panamanian military, free of charge. General Woerner referred to the practice as "sweetening the pot." He explained:

> If [at the last minute] we had a cancellation or we couldn't fill up the seats, we'd go to Panama and say, hey, we got two seats here. No cost. We were supposed to charge them, because you pro-rate costs . . . [L.G.: So why would you do that?] To sweeten the pot. You know— banking chips. I might want something someday, and hopefully, you wouldn't have to mention it; he'd know that we had provided them

with $10,000, $20,000, $50,000 worth of IMET [International Military Education Training] free. Illegal as hell, but you know, the seat would go empty. So he knew that. He was getting training beyond what he could program out of the security assistance coming to him from the United States, because we would give him empty seats. We never put it together—I mean, that would be too gross. But, you know, if on the day before the course started, we got a message saying that instead of twenty-three, [a country was sending] twenty, you wanted to fill the seat—that type of thing. [It was] taking advantage of opportunity rather than programming it.

It is likely that this practice did not begin with Manuel Noriega. The range of courses taken by Panamanian students at the SOA during the 1970s was broader than that taken by other Latin American students. Students from Panama, for example, attended forty-four different classes during this period. Yet Peruvian students only attended forty and Chilean students twenty-six, and their countries sent more students to the School. Buying the loyalty of Panamanian authorities—even corrupt dictators like Noriega—was important for maintaining strategic U.S. interests in the Panama Canal Zone.

Yet free courses were not enough to bring Noriega under U.S. control. He began to demand the removal of the School from the Panama Canal Zone, and his freewheeling relationships with the Nicaraguan Sandinistas and Fidel Castro, as well as his alleged connections to Colombian drug cartels, earned him the animosity of the United States (Scott and Marshall 1991). The U.S. military invaded Panama in 1989, removed Noriega from power and imprisoned him in the United States, where he continues to languish in a Miami jail.[27] The Panamanian Defense Forces were simultaneously dismantled, and Panama's representation at the School of the Americas ceased. Recalling Noriega many years later, General Woerner, who laid the groundwork for the U.S. invasion, was still amazed at how he had manipulated Washington policymakers. "He captured Washington," Woerner remarked. "That was part of his talent. This rinky-dink, two-bit, little guy in Panama had us jumping through hoops for years in our foreign policy. I can't believe it now, and I couldn't believe it then."

The student population of the SOA was again reconfigured in the 1980s, following the explosion of the Latin American debt crisis, the intensification of drug-related violence in Colombia, and the outbreak of war in

Central America. Mired in debt and confronted with the worst economic downturn in a generation, most countries dispatched fewer trainees to the School than in previous years. Yet total student enrollment for the decade only dropped by 3 percent from the 1970s, because three countries —Mexico, El Salvador, and Colombia—increased their enrollment severalfold. Together, their trainees represented 72 percent (9,000) of the total student population that passed through the SOA in the 1980s. The number of Mexican soldiers increased because U.S. strategists viewed Mexico as a buffer against the conflict wracking Central America, and the country was becoming an important transit point for the international drug traffic. El Salvador and Colombia both faced powerful insurgencies, and, in Colombia, well-armed drug cartels began to resist U.S. efforts to extradite their most influential figures. SOA alumni left a brutal legacy of death and destruction in El Salvador and Colombia, continuing a pattern of behavior that had become well established.[28]

Honduras figured prominently in the U.S. strategy to defeat insurgencies in El Salvador and Guatemala and to topple the Sandinista government in Nicaragua, and some eight hundred Honduran soldiers attended the School of the Americas in the 1980s. The Reagan administration transformed the country into a base of operations for a stand against what it perceived as a rising tide of communism. It authorized the Central Intelligence Agency to organize a paramilitary force of exiled, former Nicaraguan national guardsmen to conduct covert attacks against Nicaragua from bases in Honduras and Costa Rica. The administration also channeled millions of dollars worth of arms to the Honduran security forces. The military assistance strengthened right-wing elements, who, like their patrons in Washington, favored a military solution to the Central American conflicts, and who assumed the task of repressing domestic opposition. Even though Honduras did not possess a major guerrilla insurgency, military hard-liners targeted students, unionists, and peasants, as well as anyone who belonged to political parties or groups considered leftist. They also forcibly detained, tortured, and disappeared supporters of, or those believed to support, the Sandinista government or the Salvadoran guerrillas.

Within this highly militarized environment, the SOA "was not the queen on the board," according to General Woerner. Honduran officers trained in a number of other venues. They traveled to Taiwan, Guatemala, and Argentina to participate in courses on "national security" and

Figure 5. Salvadoran soa graduate Colonel Natividad de Jesús Cáceres Cabrera, the second in command of the U.S.-trained Atlacatl battalion that carried out the El Mozote massacre. Photo by Jeremy Bigwood.

Figure 6. Salvadoran soa graduate Major Roberto D'Aubuisson organized death squads and ordered the execution of Archbishop Oscar Romero. Photo by Jeremy Bigwood.

to learn about the surveillance and interrogation techniques of the Argentine military. They also participated with U.S. forces in numerous large-scale training exercises that had names such as "King's Guard," "Big Pine," and "Full Plate." Some received instruction within Honduras from Special Forces Mobile Training Teams, and an undisclosed number of Hondurans were trained and managed by U.S. intelligence agencies.

The Special Forces enjoyed a renaissance under the Reagan administration, after falling out of favor under Jimmy Carter, and the mobile training team, or MTT, became one of the primary means by which the United States instructed Latin American soldiers in Honduras and elsewhere. Supporters of the Special Forces pushed to have its work subsidized by the IMET program, and they eventually prevailed. They argued that an MTT represented a more cost-effective way to train Latin Americans in their own countries at a time when debt and recession made it difficult for governments to afford the non-IMET travel and per diem expenses for large numbers of students to train abroad. They further asserted that Special Forces in-country training was more amenable to the specific needs of particular countries than instruction at a U.S. service school in the United States. One U.S. major based in Ecuador, for example, argued that some U.S.-manufactured equipment and weapons used by Latin American militaries were obsolete and had been replaced by newer models in the U.S. military. MTTs, however, could assist local forces to operate and maintain the older equipment better than service schools in the continental United States. He claimed further that, by using the weapons available to local armed forces, the teams were less likely to stimulate appetites for expensive, unnecessary equipment among their trainees.[29]

Yet despite the increasing importance of the Special Forces, graduates of the School of the Americas continued to figure prominently in some of the most sordid actions of the Central American conflict. In Honduras, an army death squad known as Battalion 3-16 carried out most of the 184 disappearances and executions attributed to the armed forces, and its members included a number of SOA alumni. Battalion 3-16 was founded in the late 1970s by SOA alumnus and soon-to-be commander of the Honduran armed forces Gustavo Álvarez Martínez, who often selected its victims. By the late 1970s, Álvarez had "established himself as a man to watch" and the U.S. embassy considered him to be one of the Honduran military's best officers (Schulz and Schulz 1994, 85). A hard-line, anticommunist crusader, Álvarez studied police operations in Washington under the auspices of the controversial Office of Public Safety program before Congress terminated it after disclosures about the brutality of some graduates.[30] He also trained at Fort Benning, took a 1976 course on joint military operations at the School of the Americas, and enjoyed a stint at the National Military Academy in Argentina, which exposed him to the extreme right-wing thinking of the Argentine military (Schulz and Schulz 1994).

Álvarez established Battalion 3-16 with CIA and FBI assistance. According to former intelligence officer and Battalion 3-16 member Florencio Caballero, twenty-five Honduran recruits received covert intelligence training at an undisclosed location in the southwestern United States, where they were taught interrogation and surveillance techniques by U.S. instructors. These men returned to Honduras to work in Battalion 3-16, and they continued their training under the guidance of Argentine and U.S. instructors. Battalion 3-16 employed a modus operandi that resembled the tactics of the Argentinean death squads. Small groups followed victims for days or even weeks before agents driving vehicles with stolen license plates kidnapped them and took them to clandestine jails, where the disappeared were tortured, interrogated, and usually executed. Bodies were dumped along roadsides or in unmarked graves. According to Caballero, the Honduran military notified the CIA when suspected leftists were abducted, and both Caballero and Inés Consuelo Murrillo, a Battalion 3–16 victim who survived almost three months of torture, stated that U.S. personnel participated in the interrogation sessions of prisoners. An individual known only as "Mr. Mike" questioned Murrillo in her cell, and, although she was not tortured in his presence, her fragile physical condition provided a clear indication that she had been severely abused. (Human Rights Watch 1994; Cohn and Thompson 1995).

General Álvarez was not the only high-ranking member of Battalion 3-16 to receive training at the School of the Americas. Some of the death squad commanders attended multiple courses. Juan López Grijalva, for example, was an SOA student on four separate occasions between 1963 and 1975. In 1978, he went on to head the Dirección Nacional de Inteligencia (DNI), the primary operations division of the military-controlled national police force (FUSEP). FUSEP, for its part, coordinated activities between the DNI and Battalion 3-16. López Grijalva then became the armed forces' chief of intelligence in 1982, the same year that he traveled to Argentina "on intelligence matters."[31] As the head of intelligence, López Grijalva allegedly channeled orders from Álvarez to death squad operatives and oversaw their activities.[32] Generals Bali Castillo and Discua Elvir also participated in numerous SOA courses in the 1960s and 1970s before commanding Battalion 3-16 in the 1980s. Bali took Internal Defense, Joint Operations, and the SOA's flagship, Command and General Staff Officer course, and Discua trained at the School as a cadet and then returned three times to take Jungle Operations, Irregular Warfare, and Military Intelligence.

The highly fluid context in which Honduran military personnel circulated between the School of the Americas, other U.S.-based service schools, clandestine CIA-sponsored training camps, courses taught in Honduras by U.S. and Argentine instructors, the military schools of U.S. allies, and Battalion 3-16 suggests that brutal counterinsurgency tactics were disseminated from numerous U.S. or U.S.-approved venues. The School of the Americas was not unique in this regard. The circulation of military personnel between these training centers raises a number of questions. How, for example, were members of Battalion 3-16 recruited for special training by the CIA, and what was the relationship between the CIA and the School of the Americas? To what extent were contacts between future Battalion 3-16 operatives and the CIA established at the SOA and other U.S. military schools? The answers to these questions may never be known with any degree of certainty, but the courses on intelligence, irregular warfare operations, and internal defense taken by some Battalion 3-16 members at the School of the Americas clearly prepared them for what lay ahead.

Honduran officers who participated in Battalion 3-16 and rose to power in the 1980s were not only the recipients of extensive U.S. military training. They were also hard-liners who shared the geopolitical vision of the United States, a vision that was nurtured and encouraged by their military training. And they were willing to tolerate intrusive U.S. policies in exchange for large infusions of military aid that often found its way into the pockets of high-ranking commanders. Not surprisingly, as the Honduran military became integrated into a hemispheric military apparatus controlled by the United States, it grew more powerful and autonomous vis-à-vis other branches of the Honduran state and more dependent technologically, economically, and politically on the United States.[33] As a result, the United States was able to intervene in the internal disputes and rivalries that plagued the Honduran military and influence the outcomes.

Honduran officers forged close, albeit conflictive, relationships to each other in the Honduran military academy, which was created with U.S. assistance in the 1950s. Through these connections, officers forged shifting alliances that shaped patterns of mobility and informed their dealings with the United States. Promotions in the armed forces were typically the subject of internal conflict and intrigue, as aspirants contended for power and the material rewards that flowed from certain key positions within a notoriously corrupt institution. "Appetizing posts," as one individual described them, included those in the intelligence apparatus, because they allowed officials to monitor military competitors as well as to repress alleged subversives, and the United States, as we have seen, exerted enormous power in this realm. A number of agencies within the state bureaucracy, such as the customs service, the office of immigration, and the telephone company, also offered lucrative sources of personal enrichment, although the military used the pretext of "national security" to justify its control of them.

To be an SOA graduate was important for the personal and political ambitions of upwardly mobile officers, but it was never enough. School of the Americas' alumni operated in a complex field of power that extended from the military bases of Honduras to the centers of U.S. political and military power in Tegucigalpa and Washington, D.C. Within this highly unequal and contentious arena, officers trod different pathways. Some acquired wealth, power—and sometimes infamy. Those who lost internal power struggles and who never caught the attention of the United States often found themselves marginalized in dead-end positions.

Retired Colonel Juan Arias, for example, had outsized ambitions that he never realized. Arias graduated from the SOA's Command and General Staff Officer course and trained at other military institutions in the United States. He headed the FUSEP between 1975 and 1979, a position that, he claimed, placed him next in line to command the entire Honduran armed forces. But instead of great power and glory, he was passed over for promotion, given a position of lesser importance, and then packed off to Washington, D.C., where he spent a year of "gilded exile" at the Inter-American Defense College. When he returned to Honduras, he was forced into early retirement and watched the events of the 1980s unfold from his home in the hills overlooking Tegucigalpa. He maintained that his loss of influence in the army and his premature departure from it were directly related to his views on emerging U.S. policy in Honduras. He explained:

When I was the commander of the FUSEP, I was then supposed to assume the position that used to be called the Commander-in-Chief of the Armed Forces. And because this supposition existed, and the panorama of Sandinismo was emerging in Nicaragua, the U.S. ambassador used to find a way to have lunch with me in my office a couple of times a week. He asked me once what I would do as head of the Honduran armed forces given the problems that were arising in Nicaragua. I told him that I respected the constitution and the laws of the country, and that I would not allow foreign [Nicaraguan] troops into Honduras. I didn't want countries so weak, so poor, and so small to start fighting among themselves. So I wouldn't allow it. From that moment onward, the ambassador stopped coming to lunch with the same regularity. Then I started to lose power within the military. They [the military] put me over as Interior Minister. Things changed. It was no longer possible to think that I would become head of the armed forces. Other officers of lower rank were [on their way up]. The most important one was Gustavo Álvarez Martínez, who actually assumed the command of the armed forces. The history from then onward is well known.

Arias's career can hardly be described as a failure. Yet his marginalization within the Honduran armed forces came at a crucial moment in U.S.–Central American relations. It illustrates the ways in which personal ambitions and internal military power struggles conjoined with the geopolitical agendas of Washington policymakers and their ability to intrude into the internal affairs of other states. These interlocking and highly unequal encounters shaped a range of possibilities that opened up and closed down for various people within the armed forces. The training that military personnel obtained at the SOA and the international connections that they established at the School figured into these processes, and they are the subject of the next two chapters.

CHAPTER 4

Pathways to Power

There are neighborhoods in many Latin American cities where residents can point to the large, garish homes of high-ranking military officers whose salaries could not possibly purchase such ostentatious dwellings. And in the United States, especially southern Florida, many of these same officers own second homes, which they use to pursue business ventures, enjoy vacations with their families, and savor the fruits of retirement. Their wealth is often less the product of sound business investments than the outcome of decades of plunder and state-sponsored violence. That some members of the armed forces utilized their access to state power to enrich themselves, their families, and their friends and to dispossess others through corruption and brutal forms of primitive accumulation is an established fact. In Guatemala, for example, one outcome of the thirty-five-year-old civil war was "a shift in the balance of power that . . . created a new landowning elite among military officers" (Perera 1993, 11). Income polarization increased in the 1980s. The portion of national wealth controlled by the poorest 10 percent of the population dropped from 2.4 percent to 0.5 percent, while the richest 10 percent expanded their share from 40.8 percent to 46.6 percent (Vilas 1996, 470–71).

Cold war militarization greatly increased the economic power of the Latin American armed forces. Central American militaries, for example, expanded into the private sector, where they controlled lucrative business ventures that included banks, import-export operations, sweatshops, and even funeral parlors. The high command also laid claim to offices in the state sector, where, as Stanley notes for El Salvador, it "placed a high priority on misappropriating money" (1996, 70). In Guatemala, according to Schirmer, members of military intelligence received paychecks in the 1980s from agencies charged with the administration of electricity and tourism (1998, 19), and in Honduras, the military controlled the telephone company and had investments in various sectors of the economy, including construction. A Honduran SOA graduate described to me how securing a position in the state bureaucracy permitted army officers to enjoy "the kinds of benefits and enormous extras that went well beyond the salary." "Here in Honduras," he said, "you're supposed to start looking for these positions when you reach the level of lieutenant colonel or colonel. This is not just particular to Honduras but in all of Latin America."

There was, of course, nothing new about corruption, and military officers were not the only corrupt state authorities. The lack of accountability among public officials, civilian and military, reinforced an atmosphere of impunity that aggravated the plunder of public resources, the dispossession of vulnerable people, and the redistribution of wealth. What was new, however, was the military itself, as cold war militarization increased the size and the strength of the armed forces, especially armies, and opened new avenues of social, political, and economic mobility for members of the lower-middle class who increasingly filled positions in the officer corps.

The cold war blurred the distinctions between war and peace as the notion of permanent war required a larger cadre of career servicemen to confront dangers that arose seemingly from everywhere. Huge increases in U.S. financial aid, arms sales, and new training opportunities increased the size and lethal capacities of these forces. Militaries shifted from institutions mobilized periodically by the temporary conscription of civilians to permanent forces with vastly expanded opportunities for lifetime employment and professional status (Lutz 2001). To manage these larger and more bureaucratic institutions, a new group of "professional" officers arose from the middle and lower-middle class, and they displaced the amateur army leaders from the upper class who had populated the officer corps for the first half of the twentieth century.

In Colombia, for example, the military recruited the officer corps from the families of successful, small-scale agriculturalists, who tended to be politically conservative, and it avoided urban middling groups, who were perceived as more leftist than their rural counterparts (Richani 2002, 54). The opposite was true in Bolivia, where a revolution in 1952 reordered property relations in the countryside and a militant peasantry constantly confronted the state. As the army was rebuilt in the 1960s with U.S. assistance, the sons of urban petty merchants and low-level state bureaucrats assumed positions in the officer corps. These men were mestizos eager to separate themselves from any association with rural, indigenous peoples, whom they viewed as backward and ignorant.

The new officers received their educations in military academies modeled after U.S. service schools and then embarked on a series of frequently unattractive postings, service for required periods in particular ranks, and additional education at international training centers, such as the SOA. None of this attracted the sons of the upper class, but the promise of free education, career mobility, and a series of perquisites that extended well beyond the opportunities available in civilian life appealed to ambitious, social-climbing young men from middling social groups. Excelling in the military academy and forging strategic alliances to fellow officers and the United States facilitated their ascendancy through the institutional hierarchy. These experiences shaped a view of the world and a style of life that differed from those of other segments of the middle class and aligned military officers more closely with propertied elites.

Military training molded powerful beliefs about capitalist modernity, class conflict, race, and national sovereignty. Domestic military academies tantalized students with the prospects of future power and social mobility that were enhanced by the SOA experience, which exposed the Latin American armed forces to the wealth and might of the United States and sent a strong message about the benefits of allying with Uncle Sam. International military training whetted trainees' appetites for the comfortable, commodity-filled "good life" that turned on access to power, but to attain it, young military careerists had to strike a Faustian bargain: they could wield power and reap the fruits of modernity but only in exchange for learning to kill and becoming the local enforcers of U.S. policies.

We can begin to understand how military training shaped the lives and understandings of young men through the experiences of a Bolivian army officer as they unfolded, first at the Bolivian military academy during an unstable period of military dictatorship in the late 1970s, and then at the School of the Americas in the late 1980s. Although his views of the military and the School of the Americas are not typical, they illustrate some of the possibilities, ambiguities, and frustrations that young men like him confronted, as they pursued careers in the armed forces during the cold war. More broadly, his experiences demonstrate some of the ways in which military training, amid the intensified militarization of an over-heated cold war, opened avenues of social mobility for mestizos of the lower-middle class, shaped relationships within and between different militaries, molded notions about race and nation, and defined beliefs about who constituted a legitimate target and who did not. They clarify some of the ways in which imperial networks of power and privilege developed and shared understandings about national security emerged within the tension-ridden and highly unequal space of empire. They also demonstrate some of the tensions and challenges that the United States had to overcome in order to fashion Latin American militaries into extensions of its own power.

Juan Ricardo Pantoja[1] is a slight man with straight black hair and a friendly, engaging manner. In his early forties, he lives with his wife in an upscale apartment building in a large Bolivian city and currently works in the private sector. We first met in the mid-1990s, when, at the urging of a mutual friend, I sought him out to discuss compulsory military service in the Bolivian army, a topic I was investigating. We spent several evenings in local restaurants and bars talking into the early morning hours about the mandatory conscription of young, mostly indigenous, men and the ways in which the army shaped their attitudes about each other and society in general. Then we lost contact for nearly five years.

As I contemplated meeting up with him again in mid-2001, I recalled his quick wit and sharp mind. But mostly, I remembered how his incredible intellectual curiosity belied the nearly twenty years he spent in the hierarchical world of the Bolivian military, where inquisitiveness and creativity were valued less than unquestioning obedience to authority. He seemed more suited to be a sociologist or historian than an army major; indeed, after his military career was well advanced, he completed a social science degree at the state university, where he studied with some of Bolivia's most radical professors. I was therefore only mildly surprised when, after asking him to discuss his military career, he replied that "it was a big mistake." We were in his parents' home, seated around a large glass table in the dining room. It was late morning on a warm spring day, and Juan Ricardo's elderly father rummaged about the kitchen in a pair of shorts, as he prepared cups of tea and coffee for us.

His father and mother were retired schoolteachers, and their comfortable home occupied a small lot in a quiet, middle-income neighborhood.

When Juan Ricardo was growing up in the 1960s and 1970s, they could not afford many luxuries for him and his two sisters, but they managed to scrape together the money to send their only son to a private high school, where the educational standards were higher than the local public school. Yet when it came time to consider university in the mid-1970s, they faced a dilemma. The state university suffered from periodic short- and long-term closures, because the generals who had ruled Bolivia since 1964 viewed it as a hotbed of subversion, but the alternative to the university—sending their children to Mexico, Argentina, Europe, or the United States, where children of the elite studied—was out of the question. They simply could not afford the expense.

At the same time, Juan Ricardo had to confront the looming prospect of surrendering one year of his life to military service—a duty required, at least theoretically, of all able-bodied Bolivian men. Unlike poor indigenous teenagers from the countryside, who leave their families and communities to serve in distant frontier posts, Juan Ricardo and other educated urban youth had more options. In 1976, the armed forces high command announced that qualified youth could complete their service obligation by spending their senior high school year at the military academy, where rigorous study of traditional academic subjects was combined with military training. The armed forces subsidized the expenses of those that it rated highly, and, if students completed the three-year program, they became lieutenants in the army. In a society controlled by the military, a career in the armed forces opened many avenues for aspiring mestizos, and not surprisingly, competition for admission was intense. Juan Ricardo and a friend took the admission test along with 5,000 other applicants who vied for 1,200 positions in the first-year class. To the delight of his parents, he was admitted.

Juan Ricardo understood that he would receive a high school diploma at the end of the first year and that his responsibility to the military would end. What he did not understand was that if, at the end of the year, he did not pursue a military career and remain in the academy for three years, the armed forces would hold him responsible for repaying the cost of his education. "I finished the year," he told me, "and should have withdrawn from the army because I had completed my objectives, but the army laid a trap for us. It was simple: either we stayed, or we had to pay for an entire year of education. I didn't have all the information. My friend had more money than I did, so he paid and left. I couldn't pay, so I remained."

Despite feeling that he was out of place in the army, Pantoja then struggled to convince himself that he made the right decision. "I told myself that it was the most practical thing for me. My parents wouldn't have to pay for me if I stayed [in the academy] and had a career in the military, but I think that I realized that the army was not the appropriate place for me." He rationalized away his doubts and even managed for a time to enjoy his life in the military. "It was a process of self-deception," he explained. "I developed this pragmatic approach to stay and have a career. The army could offer me alternatives. The classes were very easy for me and I had no trouble getting good grades. This helped to reaffirm my decision to stay—I could continue in the army without a lot of effort. My existential conflict was outweighed by the professional conveniences." Getting good grades, however, was the easy part. Pantoja had much more difficulty adjusting to daily life in the academy, a life that was "based on terror and abuse."

Students entered a rigid world of institutionalized violence in which their lives belonged to superiors. First-year cadets underwent brutal hazing at the hands of their senior peers, and superiors punished students for even the slightest infraction with physical abuse and severe humiliation. "They would beat you with everything imaginable," he recalled, "with sticks, with belts, in your face and all over your body. Sometimes they made people jump into a pool in the middle of a freezing cold La Paz night, and others, who were evidently very sick, made people undress and stand outside naked for two or three hours during a winter night." Those on the bottom had no recourse against the brutality unleashed by those with higher rank or seniority. The result, according to Juan Ricardo, was a bubbling caldron of hatred in which savage displays of brutality were naturalized and became an accepted part of daily life. For those who managed to endure the regimen, the time came eventually when a new group of cadets arrived and the senior students did to the newcomers what was once done to them. Those who suffered the most were indigenous and provincial youth. Juan Ricardo explained:

> The military academy aggravates social and ethnic differences [L.G.: Who attends?] There are a variety [of students] from the middle class and below. Some indigenous or individuals from the provinces filter in because they have the intellectual aptitude or good academic performance. But there are those from the middle class who use their

class difference to abuse. Some of these experiences are so dramatic that cadets commit suicide. But in most cases this does not happen. What happens is an accumulation of hate that gets worked out on subordinates.

The authoritarian microcosm of the academy intensified the racism of the broader society by reproducing it in particularly brutal ways. Cadets learned lessons about obedience to authority, the rights of the powerful, and how to use force, rather than reason, to resolve conflicts. The messages were delivered through the asymmetrical relationships of race, class, and rank, and they were reinforced by a deep sense of entitlement that students acquired as they passed from derided, first-year novices, or *mostrencos* (dolts) in the language of the academy, to become senior cadets.

Students perceived themselves at the center of power, and this was especially true of Juan Ricardo's cohort, which entered the academy under the draconian rule of General Hugo Banzer (1971–78), an SOA graduate whose regime enjoyed unequivocal support from the Nixon administration. The young men studied as even harder-line dictatorships brutalized dissidents in neighboring Chile, Uruguay, and Argentina, and the Peruvian military government shifted to the right after 1975. "[Our instructors] explained the antiterrorist operations that were taking place in Argentina," explained Juan Ricardo, "and we knew about Chile and Uruguay, too. Although we did not know any details about the operations, we knew that those militaries were our allies and we were all fighting to eliminate communism from Latin America." Like the rest of the Southern Cone, open political activity in Bolivia was the exclusive right of the military, and workers' unions and peasant organizations, disorganized by repression, struggled in clandestinity to regroup.

Throughout the region, the armed forces, guided by shadowy intelligence agencies, tortured and killed thousands of people, and they forced many others to flee into exile. Although the number of deaths and disappearances in Bolivia never approached those of neighboring Chile and Argentina, the Banzer regime participated in the nortorious Operation Condor, a transnational intelligence network in which the military regimes of Argentina, Paraguay, Uruguay, Chile, and Bolivia shared information about dissident refugees in member countries, monitored their activities, and ordered their assassination and disappearance. Operation Condor was initiated by Chile and its secret police force, but it also counted on the involvement of the United States, which supported the participating intelligence services and encouraged greater participation among them.[2]

The power wielded by regional militaries led some inside and outside of the armed forces to speculate that a new era of military rule had, indeed, arrived. Describing the sentiment among his classmates, Juan Ricardo said, "We felt that we had a lot of power." In classroom exercises, he and his colleagues in the Bolivian military academy practiced attacking the crumbling building that housed the La Paz headquarters of the Bolivian Workers' Central (COB), an umbrella organization that represented Bolivian labor unions.[3] They also staged mock invasions of peasant communities—especially the militant Aymara community of Achacachi located on the flat, treeless plain known as the *altiplano*. Instructors legitimized these targets with the premise that the military was engaged in a permanent war with communist subversives, who were represented most directly by peasants and tin miners. According to Juan Ricardo:

> In our minds, Chile—a heavily armed country—did not exist [as a military target]. There was no realism. The logic of military education was organized around destroying internal adversaries. [The enemies] were the Bolivian Workers' Central and especially the miners—the most dynamic of the COB—other leaders that we identified by their speeches, and peasants. In our battle plans, the instructors indicated the physical spaces occupied by the COB and the unions. Then they would ask us [to discuss] what were our forces, who were our allies—like the Catholic Church. [They'd ask us] what has the enemy done lately? What has he said? This is how they taught us about the enemy.

The Banzer regime lasted longer than any other Bolivian government in the twentieth century, but, instead of heralding the birth of a new era, it became the first of the southern dictatorships to collapse, in July 1978. A period of economic expansion, fueled by excessive foreign credit, had come to an end, and a huge foreign debt burdened the economy. The Carter administration did not share President Nixon's enthusiasm for the Banzer regime, and it pressured the general to improve the country's human rights record. Tottering under the weight of a growing financial crisis that reached its ugly nadir only after Banzer was gone, and with only lukewarm backing from Washington, the regime could not withstand the challenge posed by a resurgent popular movement, united in its hatred of Banzer and its desire for democratization. Yet the fall of Banzer only plunged Bolivia into a prolonged period of political chaos that reflected the inability of a diverse popular movement to consolidate the political opening and the incapacity of right-wing forces to block its initiatives. Between July 1978 and July 1980, two general elections took place, five presidents held office—none as a result of electoral victory—and four military coups took place of which three were successful (Dunkerley 1984, 249).

Academy students lived the mercurial political moment with great intensity, because the school itself was an important military institution in La Paz, the Bolivian capital. Power-hungry officers, intent on seizing control of the state, cultivated the backing of the academy in the hope that the rest of the army would follow its lead. "The academy," explained Juan Ricardo, "was like an echo chamber for military coups. It was the institution where commandants, who were planning coups, went to find allies and to test their ideas. If the cadets were supportive, they became a source of legitimacy [within the military] for the coup." All of this empowered students and filled them with great feelings of self-importance. "We saw how powerful the military was and viewed [military officers] as the masters of power [*los señores del poder*]. The civilian imbeciles and idiots [according to the commandants] were the ones leading the country into disaster. The army was the only institution that could bring order." The upshot for many students was a belief that the outcome of their education was power, whether as the commander of an army battalion, the head of a government ministry, or the ruler of the country.

The megalomaniacal pursuit of power, the institutionalized violence, and the intense anticommunism that shaped the education of Bolivia's "professional" soldiers emerged from global forces—intensified worldwide militarization and the notion of "permanent war"—unleashed by the cold war. Military training was also structured locally by the volatile dynamics within the Bolivian armed forces and between the military and other sectors of society. Interservice rivalries and intense factionalism generated ever-shifting alliances and power struggles among the officer corps, and the astonishing combativeness of Bolivian peasants, miners, and other sectors of the working class posed challenges to the generals that their peers in other countries rarely confronted.

The School of the Americas played an important part in creating a mission and molding a global vision for the Bolivian armed forces, and it provided trainees with the technical expertise to deploy lethal violence. Although it was not the only U.S. or foreign institution that trained the Bolivian military, the SOA almost certainly educated more Bolivians over the course of the twentieth century than any other U.S. or foreign military training establishment. Most of Juan Ricardo's instructors had trained at the School of the Americas in the Panama Canal Zone, and, when they returned to Bolivia, they brought U.S. Army instruction manuals with them for use at the Bolivian military academy. "Usually" he recalled, "the manuals [used at the academy] that dealt with antisubversion were translations from English. In general, this was the case. The instructors brought packages of manuals that you could recognize by the color: the North American manuals were yellow." And, he added sarcastically "there were one hundred and four billion of them." As SOA-educated instructors took up teaching posts in the Bolivian military academy, entire graduating classes were shipped to the SOA every year to polish their military education. The soldiers typically took a six-month course on basic military and counterinsurgency tactics. When the students returned, members of each cohort, according to Juan Ricardo, wanted to show off their knowledge and self-importance to their junior fellows. When I asked what sorts of things they learned, he replied:

> how to tie up prisoners of war and how to torture them—techniques that you have to utilize to get them to make declarations. [For example] you don't let them sleep and then you get results. Other knowledge that they brought from the School of the Americas I remember very well. It was axiomatic among the Rangers that a dead subversive was better than a prisoner. Having a prisoner interfered with the subsequent operations, thus it's better that he is four meters underground than to have him alive.[4] [L.G.: They brought that back, too?] Yes. We must not interfere in antisubversive operations. There has to be continuity, and therefore, if we take prisoners, [the idea is to] get information from them quickly, put them four meters underground and continue the operation. [Prisoners] are a big obstacle because they generate uncertainty. If you have four prisoners, it means that you have to have a unit to guard them and another unit to feed them.

Instructions in counterinsurgency warfare and how to handle prisoners were not the only ways that the North Americans influenced Bolivian trainees at the School of the Americas. Juan Ricardo maintained that returning instructors had "internalized a culture linked to their military education that they had not developed in Bolivia." The obvious wealth of the United States Army and the contrast between the U.S.-controlled Panama Canal Zone and their impoverished homeland had a profound impact on them. One described the Zone as "like a garden." For soldiers from one of the poorest countries in Latin America, all of this provided evidence of the greater "progress" achieved by the United States. The progress, they believed, was shaped in part by the strict rules and regulations that governed every aspect of life in the U.S. Army. Such discipline was supposedly both a cause and an effect of the "order" that structured life in the Canal Zone and the United States. The North Americans had everything, or so it seemed to the Bolivians. They enjoyed a level of comfort unheard of in Bolivia: if a soldier tore his uniform, the army provided him with a new one, and the amount of food served in the SOA's mess hall made the Bolivians' eyes bulge. "[The returning soldiers] told us that you could eat like a beast at the School of the Americas," laughed Juan Ricardo.

The U.S. Army's high degree of specialization also impressed the Bolivians, whose military was not nearly as differentiated in terms of the knowledge and skills of its members. To be a specialist implied that one was special, and the ability to work with high-tech weaponry, or just modern weaponry, set the North Americans apart from their Latin American peers and students. Although Bolivian trainees resented having North American instructors whom they frequently outranked, they accorded some of them a degree of respect because of their high degree of technical proficiency. Technology—and especially the esoteric knowledge that unlocked its power—had a quasi-magical appeal for the Bolivians, and for many of these Latin Americans, U.S. Army officers seemed to go everywhere in helicopters, a symbol of their power and superiority. The conclusion that they drew, according to Juan Ricardo, was that the gringos made good allies; it was good to be on their side; and they would provide all the necessary support for the struggle against subversion. He paused and then added, "It's also better to have them as allies because they have a good intelligence system."

Yet student narratives about the comfort, power, progress, and invincibility of the U.S. military and, by extension, the United States displayed a conflicting mix of identification and admiration on one hand, and envy and contempt on the other. By the time he graduated from the military academy, Juan Ricardo had listened to three cohorts of Bolivian cadets recount stories about the SOA and life in Panama, and he felt that he knew a great deal about North Americans and their army without ever setting foot outside Bolivia. "We knew more than anybody," he insisted. "We knew how the instructors taught and how to react, and we knew which instructors were naive. There were ways of cheating [*una tecnologia de engaño*] to get better grades. There was also a lot of competition between [individuals from] different countries—a lot of nationalism. The North Americans organized this competition, and the Bolivians—real machos—understood it." Savvy students like Juan Ricardo felt prepared to engage the North Americans and beat them at their own game. Filled with nationalistic self-righteousness, they eagerly awaited the moment when they could exploit the weaknesses and idiosyncrasies of their North American patrons for maximum advantage.

Academy cadets also anticipated enthusiastically the consumer opportunities available in the Panama Canal Zone. The Zone—a center of unfettered free trade in the Americas prior to the rise of neoliberalism—offered SOA students access to a variety of cheap commodities and the possibility of acquiring the accouterments of modernity. The free trade zone, together with the Fort Gulick Post Exchange, was a veritable bazaar of electrical appliances, clothing, home furnishings, cameras, liquor, and so forth at prices considerably below those available in most Latin American countries. The dance of commodities proved particularly seductive to the Bolivian cadets whose impoverished, landlocked country with its restricted manufacturing base could only import some goods at great expense, and other merchandise was unavailable at any price. These young men seized every opportunity to shop for themselves and family members, and many also participated in a vigorous contraband trade. Because some traveled on military aircraft, it was, according to Juan Ricardo, relatively easy to smuggle goods back into the country. "[The cadets] brought everything that they could—stereo equipment and things like that. Pure contraband." The practice was winked at by the government and understood as an additional bonus for those selected to train at the SOA.

While cheap goods from Panama and the inducements of the contraband trade bolstered the social ambitions—and the everyday comforts—of SOA trainees, sex workers fortified their masculinity. SOA graduates cultivated images of themselves as manly men upon their return to

Bolivia, by regaling peers and academy students with accounts of their sexual exploits. Like a majority of their counterparts in the various armies of the Americas, many believed that access to the sexual services of local women was a basic right, and the Panama Canal Zone was presented as a place where men could indulge their sexual fantasies and escape into illusions of men-as-men. Pantoja recalled that his instructors "usually moved quickly from [accounts of] their professional experiences [at the SOA] to anecdotes about North American comfort, the prostitutes and how much they cost." Because of the enormous U.S. military presence, sex workers from a variety of countries congregated in Panamanian cities. "The brothels," explained Pantoja, "complemented other aspects of life at the SOA. [Cadets] trained from Monday to Friday, [and] Saturday and Sunday they were free. They had money so they went to the brothels [that had] black women. North Americans were there, too, and everyone was equal. The Bolivians were fascinated with black women. There are none in Bolivia and to make love with a black woman was supposedly an unforgettable experience—very exotic. It was the moment when the Bolivian military man had international contact."

The aura of almost mystic transcendentalism that surrounded the Bolivians' accounts of sexual encounters with black women emerged from a belief that you could do things with foreigners—particularly members of subordinate racial groups—that you could not do at home. Part of the allure of going abroad was the opportunity to play out sexist and racist stereotypes away from the constraints of their own society. In Panama, single men had disposable income that was unencumbered by alternative claims that would shape its use in Bolivia, and this money gave them a feeling of power and strength. It also enabled them to enter a transnational world of power and pleasure that no one at home—except for a select few—knew. As these men lived the excitement of going abroad and took part in daily training exercises at the SOA, they began to reflect on their own country in different ways.

The SOA experience aggravated long-standing domestic hatreds of "Indians" and "communists," as officers struggled to separate themselves from their own modest origins and to explain the roots of Bolivian underdevelopment to themselves. These categories were not always distinct, and their malleable boundaries insured that a wide range of personnel could be included within them. "One of the things that called my attention," said Juan Ricardo, as he described the stories told by the SOA graduates, "was the way that they talked about the United States being an organized, modern society that progressed and that there was no force that could obstruct its advance. Then they would say, 'Oh these shitty Indians. They don't allow us to progress. Nor do the communists. We have to eliminate them.'"

The extreme oppression suffered by indigenous peoples as conscripts in the Bolivian army led to their devalorization vis-à-vis the expensive, modern arms so coveted by the officer corps. Soldiers trained with these weapons at the School of the Americas and used live ammunition in many training exercises. This was not always the case in Bolivia, however, because modern weapons and ammunition were in short supply. For some officers, maintaining and protecting the limited arms available to the Bolivian army was more important than the well-being of indigenous troops, whom they viewed as eminently expendable. "They put more value in things than in people," claimed Juan Ricardo, "because Bolivia is such a miserably poor country with so few resources. This is the logic that they use with the conscripts. [When they give a gun to a conscript, they say] this is your mother. You have to love her." In this way, guns were feminized in certain contexts and imbued with emotionally laden human qualities. If not properly protected, their destruction or loss became an extremely serious matter for dispensable conscripts in an impoverished, racist army. Yet, paradoxically, in the hands of a despised ethnic group, the same guns also posed a potential threat to the privileges and the visions of capitalist modernity so cherished by the officer corps and nourished by their experiences at the School of the Americas.

After three years of imbibing the stories of repatriated soa graduates, Pantoja and his classmates awaited the day when they, too, would travel to the Panama Canal Zone and train at the soa. Yet events obliged them to wait much longer than they expected. The brutal coup d'etat of army General Luis García Meza in July 1980 slammed the door on their expectations, and for three years (1981–83) no Bolivians trained at the soa. The exceptional brutality of the coup shocked Bolivians long accustomed to life under military dictatorship and resulted, in part, from the advice of some two hundred Argentine advisors who worked with the Bolivian military. The advisors were partially drawn from the notorious Escuela Mecánica—an infamous torture center operated by the Argentine navy. Bolivia, however, has no coasts, and its small navy is for river patrols. The Argentines supervised the establishment of an intelligence network for

political control operations, and they worked closely with Bolivian Colonel Luis Arce Gómez, an soa graduate who was one of the coup architects
and the subsequent minister of the interior (Dunkerley 1984, 299). The
regime's wanton disregard for human rights and its deep involvement in
the cocaine traffic led to a fifteen-month boycott by the United States. A
broad sector of the international community also repudiated the generals
and thereby denied them legitimacy, disrupted foreign aid, and made it
virtually impossible to sustain a right-wing regime for very long. But for
the young men of Juan Ricardo's class, the coup had a more immediate
impact: it denied them the possibility of traveling to Panama and training
at the School of the Americas.

The three-year exclusion of Bolivians from the soa had less to do with
the human rights violations taking place in Bolivia than with the issue of
drug trafficking. The presumption by a segment of the armed forces that
it could use the state apparatus to manage a burgeoning cocaine trade
offended the moral sensibilities of Ronald Reagan's right-wing coalition,
and the aid boycott initiated by President Carter continued for a time
under Reagan. At the time, too, the Bolivian left presented far less a
challenge to the United States than the guerrillas who were rattling the
chains of empire in Central America. U.S. military assistance started to
flow disproportionately to Central America, as the Nicaraguan Sandinistas, the Salvadoran FMLN, and Guatemalan insurgent groups consumed the attention of U.S. policymakers.

General García Meza and his coterie of hard-line supporters would not
be denied, however. The general ordered the creation of a Bolivian military school that would train service school graduates in both conventional
warfare and counterinsurgency operations. The Escuela de Cóndores, as it
became known, opened its doors in March 1981. Recent Bolivian academy
graduates started training at the Escuela de Cóndores, rather than the
soa, and the practice continued even after the United States resumed
diplomatic relations and the military aid pipeline flowed once again. Pantoja explained: "Our discourse was that what the gringos have, we, too,
can have here. If the gringos jump a [one meter twenty centimeter] wall,
then we are going to jump one meter twenty-five centimeters. [García
Meza] got all the officers together who had special forces training from
abroad—in France, Colombia, the United States, and Argentina, and they
designed the school." Although García Meza remained in power for little

over a year, the Escuela de Cóndores became one of his enduring legacies. Despite the nationalistic rhetoric, the institution followed closely the U.S. military school model, and of the ten commandants who ran it between 1981 and 2000, nine trained at the SOA during the 1960s and 1970s. Similarly, eight of the eleven subcommandants were SOA graduates.

Juan Ricardo's academy cohort was the first class to enter the Escuela de Cóndores. "We felt a little frustrated not going to the SOA after so much expectation," he said, adding sarcastically, "We didn't have the opportunity to buy contraband or to be with black women." Juan Ricardo spent four months at the Escuela de Cóndores in 1981. Then, for the next six years, the army posted him at various locations around Bolivia, and he rose through the ranks to become a captain. In the mid-1980s, he took one of a series of periodic examinations that determined to a large degree an individual's chances for promotion, and, after scoring highly on it, he was finally offered the chance to train at the School of the Americas, which had moved from the Panama Canal Zone to Fort Benning, Georgia.

He wanted to take a course that dealt with strategy—"something more intellectual"—but his commanding officer presented him with two options: either take a basic infantry course at the SOA that repeated his training at the Escuela de Cóndores, or accept a remote posting on the Bolivian-Chilean border. The choice was obvious. Although Juan Ricardo, who was then a captain, resented the idea of repeating a course with other students whom he outranked, the SOA—indeed, any place—was better than the wind-swept army base on the dry, desolate border with Chile. Juan Ricardo rationalized his situation by deciding that he would do "absolutely nothing" in the course, but he would nevertheless take advantage of the time in the United States to learn English. "The course was of no use to me," he insisted, "and I had no interest in taking it because I was already a Cóndor, but I was interested in learning English." So, feeling slightly humiliated, he and another companion set out for a five-month stint at the School of the Americas.

"When I got to the United States," he said, "Fort Benning was nothing like all of the commentaries that I had heard about Panama. It is a town unto itself, and the School is right in the middle. I think that for the Bolivians there was a feeling of isolation. It wasn't like arriving with your entire class. I felt lonely. [Life] got very monotonous so you watch television like an idiot. There were three places to go: the training grounds, the clubs, and the px. There were clubs for sergeants and ncos where Bolivians could go and feel comfortable, and Latinas would come to these clubs . . . but at Fort Benning, [the prostitutes] spoke English, so it was more difficult than in Panama where everyone spoke Spanish."

Unlike the high-ranking and more privileged officers of the soa's flagship Command and General Staff Officers course, men like Juan Ricardo were more restricted to the post, and they had less contact with U.S. civilians. This was because their courses, which ranged from two to sixteen weeks, were shorter than the nearly year-long cgs course. The students also came without their families and lived on the base in relative isolation because they did not have the means to purchase automobiles. And the School invested less in the cultural edification of these trainees than in the more senior cgs students, who were poised to assume high-level command positions upon return to their countries. Consequently, occasional weekend outings in rental cars were the only means of interacting with the civilian world beyond Fort Benning. Juan Ricardo felt that, perhaps for these reasons, the Bolivians took training more seriously than they had in Panama.

When his soa course began, Juan Ricardo and his fellow classmates met with their instructor—a Puerto Rican sergeant—whose task was to teach them basic military skills and to prepare them to serve as leaders of infantry squads and platoons. The course covered the use of a variety of arms—the M16 rifle, grenade launcher, light and heavy machine guns, and mortars—and it addressed squad-, platoon-, and company-level combined arms tactics, such as call-for-fire procedures, communications, small-unit patrolling operations, and land navigation techniques. Salvadoran lieutenants and second lieutenants, fresh from the war-torn battlefield of El Salvador, made up the largest group of students in the class; indeed, Salvadorans predominated at the soa in the 1980s as civil war engulfed their country.

For Juan Ricardo, the contact with students from other countries was enlightening but not entirely satisfactory. "I met a lot of Salvadorans," he said. "They were really ignorant, brutally ignorant. But luckily I made friends with two Costa Rican policemen—one was a right-winger and the other shared my views—and together we reflected on the culture of Central American countries. The Ticos [Costa Ricans] made a lot of jokes about the enormous ignorance of the Salvadoran military." When I asked him to elaborate more on his impressions of the Salvadorans, he replied:

The majority of students were Salvadoran, and they were training to return to combat in El Salvador. Those guys thought about three things: first, they wanted to train themselves well; second, they wanted to buy pick-up trucks and drive them back to El Salvador. When I finished class at the end of the day and went to the library, they would go out and look for cheap pick-ups to buy. And third, they had a lot of relatives [who they wanted to see] in the United States— especially Washington, and it was not the first time that they had been to the United States. They admired the United States in the same way as the Bolivians who trained at the School of the Americas.

Because of their close contact with the United States, the Salvadorans, according to Juan Ricardo, could maneuver within the cultural milieu of Hispanic America, and they had excellent relationships with the primarily Puerto Rican sergeants and instructors who, he insisted, ran the SOA. But the "comfortable relationship" that he described between the Salvadorans and the SOA instructors went beyond common cultural understandings shared between two minority groups in the United States, and this be- came evident as he continued:

Their relation with the sergeants was excellent. They spoke the same language, and they had the same customs. In fact, instructors had participated in combat, in support of the forces in El Salvador. They were [all] like a big family. [L.G.: Did the instructors participate in combat or were they in El Salvador as advisors?] They fought and they were intelligence advisors. The [Salvadoran] battalions had a lot of support from the North Americans—North Americans fought with them. [At the SOA] the training was really for the Salvadorans. It wasn't for us [the Bolivians]. We were a pretext. The instructors didn't pay attention to us. I was happy, but my [Bolivian] companion got depressed. The School of the Americas was for Salvadorans.

Being stuck in a course with Salvadoran lieutenants and second lieu- tenants whom he outranked did not sit well with Captain Pantoja, and revisiting material that he well understood with a sergeant instructor from Puerto Rico, whom he described as a lead brute [*bruto de plomo*], further offended his professional sensibilities and notion of military hier- archy. Not surprisingly, Juan Ricardo entered the course with a large chip on his shoulder, and he defended his wounded ego by behaving like an

incorrigible smartass. The result was a negative evaluation from the director of training. As Juan Ricardo struggled with his personal feelings, he
had to contend with the ideological bromides and simplistic analyses that
accompanied the process of political education at the SOA. He explained,
"[The sergeants] said that all the communists in Latin America were
trained in Cuba and that they hated their countries. Those of us who were
at Fort Benning were going to become the leaders of our countries. We all
had to unite against communism. I questioned the simplicity of all of this.
I was very imprudent. The sergeants just repeated what they learned from
their own instructors." When I asked him to describe the course in more
detail, he continued:

> For example, there was a section of the course called civic action. It
> was one of the moments when the [anticommunist] doctrine really
> came out. They taught you that when you enter a village and make
> contact with the population, you have to make sure there are no
> communists. [L.G.: How do you do this?] They never said. You never
> trust anybody. You never enter a home and accept a plate of food
> because a communist might have poisoned it. These people are not
> going to be free because of their Marxist indoctrination. I had an
> argument with one of the sergeants. I asked him to explain Marxist
> doctrine but he couldn't, so I explained it to him. It was great. I had
> already taken a year of [social science] classes at the university [in La
> Paz]. They [the sergeants] know only formulas. The objective is to
> homogenize the education of the SOA students.

The challenges that he posed to instructors and his defiant behavior
caught up with Juan Ricardo when he returned to Bolivia. He was reassigned to a base on the Chilean border that he had initially avoided by
attending the School of the Americas. There were very few amenities, and
nothing happened. Pantoja believed that he was being punished less for his
behavior at the SOA than for continuing to pursue a university degree in
the social sciences. It was bad enough that he had chosen to enter the
university, which was a hotbed of opposition to the military, but his field of
study particularly angered his superiors, who viewed it as a "red" discipline. Pantoja used the time on the Chilean border to write his thesis, but
when he was finally posted in La Paz, his difficulties in the army persisted.
"On many occasions," he said, "I had problems with other officers. They
would get drunk and tell me that I was not like them and that I disrupted

the *esprit de corps* that they conceptualized in a particular way. I didn't let it get to me because I knew what kind of people they were and the complexes that they had with respect to knowledge and criticism."

Although Juan Ricardo remained in the army for three more years, he eventually retired in his late thirties and pursued a career in the private sector. His experiences cannot be generalized to everyone who passed through the School of the Americas, but they illustrate an aspect of the harsh reality of the cold war and war preparation that is normally shrouded in silence and official propaganda. They demonstrate some of the ways that shifting hierarchies of power based on race, class, gender, and nation defined relations of domination and subordination and justified different forms of violence. Cold war militarization limited the educational and career options for lower-middle class mestizos, and it drew many into the armed forces, which offered one of the few channels for professional advancement in a society dominated by the military. The young men entered the armed forces at a time when the United States was exerting greater control over the security forces and transforming them into instruments of its own power. Consequently, newly minted officers became both the local level enforcers of the domestic status quo as well as the servants of a far-flung empire headquartered in Washington, D.C.

The enticements of training abroad encouraged them to collude with the imperial project of the United States. The chance to participate in an exotic world of sex and consumer commodities tantalized young academy trainees years before they actually traveled to the School of the Americas, and the modern weaponry and technological sophistication of the hemisphere's most powerful military fascinated the Bolivians and reinforced their claim to monopolize the use of lethal violence at home. Access to this transnational world of power and privilege was tied to their allegiance to the political goals of the United States and to participation in the armed forces. Yet the emergence of a hemispheric military apparatus in which the Bolivians participated as junior partners aggravated processes of social, economic, and political differentiation locally. Trainees used experiences abroad to separate themselves from their own humble social origins in Bolivia. Perhaps more importantly, however, arming and training a segment of the lower-middle class, providing it with channels of upward mobility, and assigning it the task of repressing those who challenged the regressive political and economic policies of the state aggravated long-standing forms of race and class oppression in Bolivia.

✛

CHAPTER 5

Strategic Alliances

"It would be a mistake not to maintain the School of the Americas," commented retired Colombian General Alberto González Herrera. "Going to the School is a stimulus for certain officers who have merit. Their families go, and they are paid in dollars, which is an advantage. And the School connects people from other countries in the Americas." González was referring to the SOA's flagship Command and General Staff Officer course (CGS), the year-long class for upwardly mobile, midlevel officers. "I don't believe that what the boys learn or don't learn is particularly important," he continued. "They learn many things, but that is really of secondary importance. The relations that they establish with others are at bottom the most important. . . . The School also permits the United States to have the future leaders of the [Latin American] armed forces in its hands. If [U.S. commanders] don't think about that, they are making a mistake."

González knew his subject well. He attended the School of the Americas on several occasions and took the Command and General Staff Officers course in 1973. "When I was in the service," he said, "there were [people who became] colonels and generals who took the course with me. Afterwards, when I had a command and traveled to, say, Argentina or Brazil, there was the person from the course. This had more political and strategic importance than what actually got taught in the class." As a senior officer, González also influenced the selection of Colombian soldiers who trained at the SOA, and he did so with an eye to building connections between the Colombian armed forces and other militaries of the Americas, particularly the United States.

González and I were sitting in comfortable leather chairs around a coffee table in the plush office of his colleague General Néstor Ramírez Mejía, the second-in-command of the Colombian armed forces. It was August 2001, and the forty-year Colombian civil war was intensifying with the infusion of millions of dollars of U.S. military aid for what Washington policymakers then described as an antidrug war. Ramírez, a distinguished 1985 CGS graduate, was playing a key part in prosecuting the U.S.-financed war, which, on the ground, focused less on drug traffickers than on leftist guerrillas struggling to overthrow the government. He, like other high-ranking Colombian graduates of the School's CGS course, barely mentioned drugs or drug traffickers in their conversations with me. Yet he had a lot to say about the guerrillas, especially the Fuerzas Armadas Revolucionarias de Colombia (FARC), the oldest and largest insurgent movement in the country. Ramírez, González, and their colleagues reviled the guerrillas, and the country's abysmal human rights record testified to the armed forces' brutality in combating the insurgents. According to numerous national and international human rights organizations, state security forces and allied paramilitaries were responsible for most of the disappearances, massacres, and extrajudicial executions that made Colombia one of the most violent places on earth (see chapter 7).

Before escorting González and me into his inner office and departing for a meeting, Ramírez held forth beside a portrait of Jesus for nearly forty-five minutes about the war, human rights organizations, and guerrillas. He explained that official military policy constrained his ability to speak out against human rights organizations that were, he claimed, in league with the guerrillas, but as he warmed to the topic, official policy did not seem to restrict the remarks that he made to us in the privacy of his office. Ramírez was upset about the recent treatment of a fellow officer who had been led away in handcuffs after Colombian law enforcement officials entered his home. The officer, General Rito Alejo del Río, who attended the SOA as a cadet in 1967, had proven ties to paramilitary death squads over many years, and human rights organizations had long demanded his arrest. His recent detention took place amid mounting international pressure on the Colombian government to sever its ties to the paramilitaries. The arrest caused consternation among men like General Ramírez, because del Río was the highest-ranking official ever detained for paramilitary involvement. It was, however, no surprise when del Río was released less than a week after the arrest, and the human rights coordinator in the attorney general's office was ordered to resign for detaining him. Nevertheless, Ramírez's indignation had still not subsided.

He blamed del Río's difficulties on human rights organizations. Several Jesuit-affiliated groups in Colombia were, he insisted, linked to the guerrillas, and he maintained that international bodies, like Human Rights Watch/Americas, were directed by "Marxists." In Ramírez's estimation, General del Río was a fine officer who had "pacified" Urabá, a particularly conflicted region of the country, and the humiliating circumstances of his arrest were unconscionable. Ramírez, however, had no problems with the ruthless violence unleashed by paramilitaries after del Río took command of a battalion near the town of Apartadó, nor did he mention the repeated and futile requests for protection that besieged civilians made to the army.

Men like Ramírez and González play key parts in the implementation of U.S. policies in Colombia and elsewhere, and the CGS course is one place where the U.S. Army cultivates ties to them. The course provides a strategic vantage point from which to explore where officers have been in their careers and where they are likely to go, as well as to consider the relationships that they establish with each other along the way. Taking the course at a military training center—domestic or foreign—is a requirement for midlevel officers who seek promotion to colonel in the militaries of the Americas, and admittance to the CGS class makes or breaks the careers of aspirants to the senior ranks of the armed forces. The course is the longest and most prestigious class offered by the SOA and brings students and their families together with the U.S. military for a year. The SOA sets its sights not only on the military training of CGS students and their incorporation into a U.S.-dominated military apparatus. It also concerns itself with the cultural education of trainees and their accompanying dependents, as it seeks to break down nationalist sentiments, create new, transnational social networks of officers, and spread the values that it associates with the American way of life to the future military leaders of Latin America. The class thus provides an important venue for understanding the formation of the senior officer corps and their ties to each other, as they circulate around the Western Hemisphere. The relationships forged between U.S. and Latin American officers help to promote U.S. policies locally, and they serve as important channels through which the U.S. military penetrates Latin American security forces and exerts control over them.

The Command and General Staff Officer Course

Some nine hundred Latin American officers took the CGS class at the School of the Americas between 1955, when it was first offered in Panama, and 1996. Many trainees participated in the course, returned to their countries, and served out unremarkable careers that attracted little public attention, but others subsequently rose to prominent positions in their militaries. Some also achieved considerable notoriety. For example, Chilean General Augusto Lutz, who graduated in 1966, participated in the U.S.-backed coup d'etat that overthrew Salvador Allende in 1973. Guatemalan Colonel Julio Alpírez graduated in 1989 and became an important CIA contact within the Guatemalan Army. While on the CIA payroll, he supervised the prolonged torture and execution of guerrilla leader Efraín Bámaca. More recently, the commander-in-chief of the Venezuelan Army, Efraín Vásquez—a 1988 CGS graduate—backed a 2002 coup d'etat that temporarily deposed President Hugo Chávez and that had heavy overtones of U.S. involvement.

Five countries—Venezuela, the Dominican Republic, Honduras, Guatemala, and Bolivia in order of importance—have sent the most students to the Command and General Staff Officer Course. Six hundred and thirteen officers from these states account for 68 percent of the total. In contrast, only 6 percent of the trainees come from the southern cone countries of Argentina, Chile, Uruguay, and Paraguay, and Brazil has only sent students to the CGS course since the late 1990s. Colombia, El Salvador, Mexico, and Peru occupy a middle ground. They account for 22 percent of total course participants and have dispatched between twenty-seven and sixty-seven students apiece over the years.[1]

Because of the gate-keeping function of the CGS class, admission is highly selective. Only the rising stars of the Latin American armed forces gain entry into this elite body of young officers. Most students first take the course at home, where they typically graduate at the top of the class and then, because of their academic performance, distinction in combat, or the favor of superiors, are rewarded with the opportunity to enroll at the School of the Americas and take it again in the United States. The class is therefore less a new training opportunity than a gift for past achievements to officers already identified as potential leaders. According to the SOA commandant, Latin Americans come to the School because they want "to play in the big leagues" and "to rub shoulders with the most powerful military in the world." Enrolling in the soa for these officers is a bit like attending an elite finishing school that reinforces beliefs about the superior qualities of select students and that equips them with the personal connections, advanced military skills, and cultural understandings required for participating in a far-flung, imperial military force.

The cgs course is modeled after a class by the same name at Fort Levenworth, Kansas, the U.S. Army's premier officer-training center, where some one thousand Latin Americans took a nonresident version of the class over the course of the twentieth century (Madden 2000). The Latin Americans represented 18 percent of the foreign trainees and included men like Guatemalan General Héctor Gramajo, the notorious architect of the Guatemalan military's counterinsurgency campaign. Classes at Fort Levenworth are conducted entirely in English, and although Latin American trainees might prefer to attend Fort Levenworth, many are excluded by their inability to speak English and are channeled to the soa instead.

Latin Americans, however, are not the only trainees to attend the cgs course at the School of the Americas. Unlike other soa offerings, the cgs class contains a sizable contingent of U.S. officers. The participation of U.S. students has increased since the School moved to Fort Benning and the class received accreditation by the army as the equivalent of the Fort Levenworth version. U.S. officers now fill approximately twenty positions in a course that contains about sixty students every year. Aspiring majors are admitted on the basis of test results, detailed performance evaluations that assess past jobs and speculate about future capabilities, and informal networks of communication that bring ambitious officers to the attention of important superiors. They compete for 1,150 openings at Fort Levenworth and roughly 75 positions at other military installations, including the School of the Americas. Those who are not accepted anywhere either take the course on a correspondence basis or give up their hopes of becoming a colonel. Commenting on the strategic importance of having U.S. soldiers in the class at the soa, former commandant William De Paolo (1989–91) explains that "the whole idea was to have officers who were going to be working in Latin America, either as foreign area officers or Special Forces officers, or sergeants for that matter. It established a kind of rapport beforehand, and even if they don't run into the same people, it gives these guys a sense of what to expect and what customs and cultural differences are. And [it's a chance] to polish their language skills."

Yet while the Latin Americans in the cgs course are being groomed for future leadership positions in their respective militaries, the U.S. students at the School of the Americas—particularly the Euro-Americans—feel that they have been relegated to a backwater that threatens to drown their careers. They appreciate the opportunity to take the CGS course, but most would prefer to do so at Fort Levenworth, which is securely identified with the Euro-American mainstream of the army and ostensibly opens more opportunities for career-enhancing European postings in combat arms assignments that have command duty. Fort Levenworth places U.S. students in a special resident section of the course and requires that they possess a top secret security clearance in order to study material that is not part of the SOA curriculum. It segregates foreign students in a nonresident section that has more lenient security restrictions.

The ambivalence of U.S. officers about the SOA is reflected in administrators' complaints that they carry "a chip on their shoulders" and "believe that they are missing something that they could get elsewhere." SOA Commandant Glen Weidner was a 1986 CGS graduate and hoped for a European assignment when he finished. The army, however, sent him to Panama, and his subsequent career remained focused on Latin America. When I asked him to discuss the relative importance of a Latin American posting for advancement in the armed forces, he became exasperated and replied that "at some point you just have to grow up."

Like other SOA classes, the CGS course uses Spanish translations of manuals developed and used at Fort Levenworth, but officials censor sensitive subjects in these manuals because of the political controversy that envelops the School. Former commandant Colonel Roy Trumble (1995–99) maintains that the curriculum is "lighter, sweeter, and kinder than the same curriculum that is taught elsewhere" and insists that the School gives Latin American students "the vanilla flavor . . . to insure that they don't think up any bad things from us." Trumble, who took the CGS course at Fort Levenworth, asserts that "the Command and General Staff course that I attended at Fort Levenworth had more harsh things. It had more open discussion of military intelligence, special operations, you name it." When asked for a specific example, he continues:

> There was a seminar in which writings from four or five people were put out there. One advocated civilian control, and it talked about how guerrilla warfare had to be harsh, and how suspension of civil rights was necessary as a rule, not as an exception. It talked about very

controversial programs in Vietnam, like the strategic hamlet program or the Phoenix program . . . as being a necessity. So we pulled
that out and don't give that to our students. We just don't deal with
controversial issues that would have some sort of political significance
in terms of human rights, whereas Fort Benning may not be so sensitive to it. Fort Levenworth and the U.S. Army War College may not
be as sensitive, either.

Since the eruption of the human rights controversy at the SOA, officials
have adopted various strategies for dealing with critics, including water-
ing down course content and renaming controversial classes (see chapter
8). This apparent concern for human rights has not affected training
schools far from the center of public criticism, and Trumble does not
comment on the implications of "harsh" instruction for the behavior of
U.S. soldiers who graduate from these schools.[2]

The CGS course is divided into a series of instructional blocks that
review key areas of expertise required of field commanders and brigade-
and division-level staff officers. Over the years, the categorization of the
material has varied in accord with political developments and changing
military doctrine. In the 1964 SOA catalog, for example, the course subjects
are broken down into fifteen areas in which the "fundamentals of military intelligence," "fundamentals of counter-insurgency," and "counterinsurgency operations" account for 189 instructional hours, or 15 percent
of the course offerings (USARSA, 1964), but by 2000 these categories vanish
or appear under other divisions.[3] A student handout delineates tactics,
strategy, civil-military affairs, and a term paper on a selected theme as the
major topics of study that count for 60 percent of the course grade.
Despite the variation, there has been a considerable degree of continuity
in the class. A large instructional block always focuses on planning, logistics, leadership, and resource management, which are some of the major
duties and responsibilities of general staff officers, and the class includes
instruction in a mixture of conventional and so-called unconventional
warfare techniques.

Considerable preoccupation with conventional warfare, such as large-
scale joint and combined operations at the brigade, division, and corps
levels, demonstrates the strength and organizational capacity of the U.S.
military to the Latin American trainees, and it enables Latin American
officers to experience U.S. power vicariously through mock exercises and

class projects. In reality these operations are frequently beyond the scope of many Latin American militaries that are not large enough to possess divisions and corps. Armies are typically the largest segment of these armed forces, and the absence, or the antiquated nature, of other service branches limits the development of combined exercises. Yet the emphasis on large-scale conventional warfare in the CGS class locates Latin American militaries within the hierarchy of hemispheric military power, and, as the commandant suggests, it gives midlevel officers a chance to "play ball in the big leagues." Students' affiliation with a major-league military is further developed through visits to important U.S. domestic and international military installations during the year.

On a more practical level, the CGS course teaches students how to integrate counterinsurgency operations into broader strategic plans. It has always paid attention to "internal security" through an examination of activities that have been alternately labeled counterinsurgency operations, low-intensity warfare, or operations-other-than-war and that include intelligence gathering, psychological operations, commando techniques, civic action, and small-unit patrols. CGS students understand the basics of these activities from previous military training, and many have participated in them as platoon and company commanders earlier in their careers.

The Command and General Staff Officer course is therefore not only an important career step into the military elite. It is also a form of management training that reinforces the deep class differences that divide all militaries and that bind Latin American graduates more closely to their patrons in the United States. Most of the U.S. and all of the Latin American trainees are graduates of police and military academies. In Latin America, these instructional centers are based on U.S. military school models, and, as we saw in chapter 4, they breed intense nationalist sentiments. International training centers like the SOA must manage these nationalist sentiments created by soldiers' prior training. The U.S. military has done much to create these nationalisms by channeling billions of dollars of aid to assist countries' battles against alleged internal enemies. International training centers also try to incorporate and subordinate upwardly mobile trainees into a broader vision of imperial solidarity. The CGS course is an important arena where the imperial arrogance of the United States and the nationalist pretensions of Latin American militaries collide.

The CGS Class of 2000

In early January 2000, the School of the Americas sprang to life after an end-of-the-year holiday lull. The cGs students began arriving in Georgia from points across Latin America and the United States, and they set to work equipping households, enrolling children in school, obtaining valid drivers' licenses, registering for class, and engaging in sundry other tasks. The course contained sixty-two students, making it one of the largest groups in recent years. Nearly half (29) were U.S. soldiers from various branches of the armed forces. Most came from the army, but two air force majors, four special forces officers, and four national guardsmen counted themselves among the group. The remaining thirty-three officers were Latin Americans who represented ten countries: Argentina (1), Bolivia (10), Colombia (2), the Dominican Republic (4), El Salvador (2), Guatemala (2), Honduras (1), Mexico (2), Peru (2), and Venezuela (7). Like their U.S. classmates, most were from the army, but a ten-member Bolivian contingent arrived with seven policemen who belonged to militarized police forces created in recent years to confront militant peasant coca growers.

The cGs officers displayed a broad range of expertise. U.S. Major Carmen Estrella—an intelligence officer—specialized in psychological operations (PSYOP) and was a graduate of Fort Bragg's training program for PSYOP specialists. A veteran of the 1991 Gulf War, Estrella suffered from the health problems associated with "Gulf War syndrome," but she received little sympathy from many of her superiors who claimed that such a syndrome did not exist. When I asked her what it was like to be the only woman in the course, she said that it was not a new experience for her. "I just tell them to treat me like one of the boys," she said, "but, of course, they don't."

Another U.S. major, Rafael Torres, also specialized in intelligence. Torres said that he worked in the Pentagon and planned to return there at the end of the year. In the 1980s, he operated inside Nicaragua with the Contras. He then shifted to El Salvador, where, from 1989 to 1992, he advised the Salvadoran military in its war against the guerrilla insurgents of the Farabundo Martí National Liberation Front. He was based in the town of San Miguel and responsible for intelligence and counterintelligence activities for the entire region east of the Lempa River. These duties included instructing the Salvadoran military on how to interrogate and reinterrogate prisoners.

Major Thomas Danek was one of two U.S. Air Force officers in the course. Thirty-something, baby-faced, and chubby, Danek was a pilot who had just surfaced from the air force's so-called Black World of highly secret undercover operations. Every day for four years he had reported to work, dressed as a civilian, at a downtown Las Vegas office building rented by the air force. His superiors prohibited him from discussing his job with anyone, even his wife, and if asked about his work, he could only reply that it involved "testing systems." Danek clearly reveled in the secretive nature of his work, and he garnered a certain amount of self-importance by letting others know about it and the multiple layers of cover that had shrouded his activities. In a display of bravado that highlighted pervasive interservice rivalries, he scoffed about his friends in the Rangers who liked to think of their work as highly clandestine. "Customers," he claimed, had come to his office "because they did not want to deal with the Rangers."

U.S. Major Paul Dececco was an infantry officer and a graduate of the army's Foreign Area Officers Program from which graduates derived the unofficial sobriquet FEO, or ugly, in Spanish. FEOS not only carried their branch designation—infantry, in Dececco's case—but also specialized in an area of the world to prepare themselves for political-military assignments. The U.S. Army encouraged FEO trainees to study at a military academy in the world region of their specialization, and Dececco had taken the Guatemalan military's six-month "Advanced Course" in which he studied with some sixty senior Guatemalan captains and majors. These classmates did not impress him. "They plagiarized like crazy," he said. Yet the course served a number of useful purposes beyond military training. It was, he pointed out, an excellent way to make contact with Guatemalan officials early in their careers, and it gave him the opportunity to observe the professional development of the Guatemalan officer corps. Although Dececco seemed put off by the Guatemalans' rampant plagiarism, studying with an army that had one of the world's worst human rights records did not appear to bother him.

Dominican Colonel Víctor Rodríguez was the lone colonel and the highest-ranking officer in the course. Rodríguez came to the SOA because he needed to be out of the Dominican Republic during a period of political transition. Rodríguez wanted to minimize his association with the current government, in case it lost an upcoming election, and strengthen his chances of working with whatever party was victorious. The CGS course served this purpose well by enabling him to disappear for a year and thus avoid involvement, or the appearance of involvement, with the political contenders back home. It was also, he assured me, more prestigious to attend the soa's Command and General Staff Officer course than its equivalent in the Dominican Republic. Because of his rank, Rodríguez became the official spokesperson for the cgs students and represented them during official events and ceremonies.

The Bolivian group contained men such as Major Julio Miranda, a member of a militarized police force who operated on the front lines of the U.S. drug war in the coca-growing region of Chapare (see chapter 7). Army Major Oscar Pacello was another member of the Bolivian group. Pacello had little respect for policemen like Miranda, who he felt were unqualified for the cgs course, and his admiration of the U.S. military came mixed with a certain frustration and anger at its inability to understand what was going on in Latin America—particularly in conflict-filled countries like Bolivia—from perspectives other than its own.

As the alumni of military and police academies, U.S. and Latin American students are always associated with their particular class. In Latin America, academy graduates become lifetime colleagues who are promoted together—at least theoretically—and whose career pathways constantly intersect. At particular junctures, all students participate in a progression of postacademy educational and training experiences that shape their movement up the military hierarchy. The cgs course is one of these. The trainees are ten to fifteen years beyond the military academy, but they have taken additional courses in their home countries, at various services schools in the United States, including the soa, and in third countries, such as Taiwan, Argentina, Spain, and Chile, to acquire new skills and perspectives. Among the students of the 2000 class, for example, Bolivian Major Oscar Pacello took a three-month course on military intelligence at the soa in 1994, when he was a captain, and graduated with honors. He then returned a year later to participate in Psychological Operations. His cgs classmate, Dominican Major Luis Rafael Abréu, studied basic infantry tactics at the School in 1985. These educational interludes are typically interspersed with command, staff, and teaching assignments in the student's home country.

As the class of 2000 registered, made the required protocol visits to the commandant's office, and attended various orientation sessions, students sized each other up, paying close attention to the pins, patches, and insignia affixed to the uniforms of their classmates. The accessories offered an important road map to each individual's career by broadcasting key information about rank and status and displaying evidence of the accomplishments of students who would contend for end-of-the-year prizes and honors. They also helped the men form opinions of each other at a time when the cliques, animosities, and friendships that would crosscut national alliances and shape group dynamics over the next eleven months had not yet emerged.

soa administrators selected U.S. student sponsors from among the incoming cgs trainees to help the Latin Americans adjust to life in the United States and to establish relationships with them. Because of his experience in Guatemala, Major Dececco, for example, became the official cgs student sponsor of Guatemalan helicopter pilot Lieutenant Colonel Octavo Galindo, who was one of two Guatemalan students participating in the course. The other Guatemalan, Major Luis Juárez, was assigned to National Guardman Major Michael Montoya. Both Galindo and Juárez had already taken the cgs course in Guatemala and were veterans of Guatemala's civil war. The short, mustachioed Juárez, who would eventually graduate first in the class, belonged to an elite, army special forces unit known as the "Kaibiles."[4] Galindo's and Juarez's sponsors—Decceco and Montoya—were expected to assist them and their families in getting established in Colombus and to "model U.S. military values" (e.g., selfless service, duty, honor, integrity, and respect) for them.

The responsibilities of the student sponsors were thus practical and ideological, and Colonel Weidner remembered the duties from his days in the School's cgs course. Weidner recalled the time in late 1985 when he and his U.S. classmates met with the School's commandant and the cgs course director, as a new cohort of students from Latin America arrived for the 1986 class. The officials wanted Weidner and his peers to understand the task that lay before them. According to Weidner, "The commandant at the time told me and my seven U.S. classmates in the [cgs] course that the reason we were there was to transfer values to Latin American officers. They will go home not only knowing our doctrine but also understanding our role in a democratic society. And that kind of impressed me," he said. "There was absolutely no doubt in anybody's mind that this was a redemptive effort for the Latin Americans."

The army's attempts to "redeem" Latin Americans is continued by U.S. host families who complement the work of assistance, trouble-shooting, and indoctrination started by the student sponsors. School officials select host families from a roster of approximately sixty civilians and retired military officers who reside in the greater Columbus area, and they then assign one family to each Latin American student and his accompanying dependents. Former SOA commandant José ("Joe") Feliciano has hosted Latin American families on a number of occasions. So, too, has Cory Loomis, a retired army officer and past president of the SOA Host Family Program. Loomis estimates that he has worked with nine CGS students or Latin American guest instructors and their families over a twelve-year period, and his background and demeanor epitomize many of the characteristics desired of a host family mentor. Married with children, Loomis lives in suburban Columbus and devotes much of his free time to a local Boy Scout troop. He is a native Southerner who prides himself on the ability to speak standard English, "Southern" English, and the profane vernacular of the army. He also speaks Spanish, which he studied for four years at West Point and then perfected during various assignments at the School of the Americas in Panama and Fort Benning between 1978 and 1986. During these postings, he served as the company commander, worked as the chief of tactics in the department of combat operations, and ran the School's public affairs office. By the time he retired from the army, he was very familiar with the SOA and its students.

Reflecting on his participation in the Host Family Program, Loomis said that sometimes his relationship with the students and their families "clicked and sometimes it did not." Although his level of involvement fluctuated in accord with competing demands on his time, Loomis liked to show Latin American families "some of the things that are available in the local community," including an annual steeplechase, Atlanta Braves baseball games, and backyard barbecues at home with his family. On a number of occasions, he helped students overcome problems that arose in the course of their stay in Georgia. One man, for example, purchased a new car and unwittingly signed up for nearly a thousand dollars of additional charges above the purchase price. Loomis talked to the dealer and convinced him to drop the bulk of the expenses. Another student had to move onto the base after the devaluation of his national currency, but when he tried to break the lease on his apartment, the landlord refused to refund the deposit. Loomis came to the rescue. He intervened with the apartment manager, explained the situation, and retrieved the deposit.

Through such quotidian pleasures and aggravations, Loomis and other host family representatives introduced the Latin Americans to a slice of middle-class, Southern life, generalized the experience to "America," and helped them to make sense of it. They also humanized it as friendly, helpful, and leisure-oriented.

The cultural education of the CGS class extends to wives and children. Army rhetoric proclaims that "the military is a family," and since the Vietnam War, the U.S. military has become more concerned with the ways that family problems can affect the retention and readiness of its forces (Enloe 2000). If women find their own lives satisfying, they are more likely to encourage their husbands to remain in the military. If Latin American wives are allowed to participate with their husbands in the "modern" world of the United States, families are more likely to remain together, enabling husband and wife to serve as cosmopolitan representatives of their national militaries. Living in the United States also allows CGS families to accrue cultural capital that distinguishes them from military and civilian peers at home. Such capital includes the acquisition of cheap commodities and visits to Disney World, the Smithsonian Air and Space Museum, and upscale shopping malls, all of which locate them squarely among the prosperous, global citizens of the modern world. Even though in the United States some of the trainees and their relatives live in Columbus's poor minority neighborhoods or in less than glamorous housing in Fort Benning, the ability to acquire commodities, visit key tourist destinations, have their children learn English, and simply associate with a modern superpower facilitates their identification with a cosmopolitan vision of modernity once they return home (cf. O'Dougherty 2002). International travel and living have long been tools in the class struggles of diverse peoples, and they are also part of the process by which the United States binds up-and-coming officers and their wives to its vision of the world.

For the children of CGS families, living in the United States provided the opportunity to attend free public schools that were comparable in many cases to the private schools in which they studied at home. Even though the quality of education in the Columbus suburbs and the poor neighborhoods around Fort Benning varied, all the schools provided foreign students with the opportunity to learn English, and English fluency was an important tool that families used to intensify status distinctions at home and to secure the class standing of the next generation. It identified the speaker as cosmopolitan and knowledgeable, and it opened a variety of opportunities that were unavailable to monolingual Spanish speakers.

Patricia,[5] an eight-year-old Bolivian who accompanied her parents to the SOA in the late 1990s, picked up English quickly in the Fort Benning elementary school she attended, and she entered a bilingual private school after she returned to Bolivia. When I visited the family's Cocha-bamba apartment to interview her father in 2000, both father and mother quickly called my attention to their daughter's fluency in English and encouraged the little girl to speak with me. They were clearly very proud of her. Patricia told me in nearly perfect English that she had a lot of fun in the United States and did not want to be in Bolivia. When I asked her to explain, she replied that all the poor people on the streets of Cochabamba bothered her, and she did not like to see them every day. Life in the United States had been preferable because one did not have to look at poor people all the time. Her mother agreed and explained why the time in the United States had been so exceptional. "It gave the family an opportunity to know a new way of life," she said, "a way of life that is modern and oriented toward the future, even in terms of how to raise children." For Patricia, learning English, residing in the United States, and the inter-pretation of these experiences provided by her parents reinforced a sense of class distinctiveness and cosmopolitan privilege that set her apart from others in her hometown.

The SOA commandant's wife, Marcia Weidner, devotes much of her time to maintaining morale and promoting social cohesiveness among the wives of CGS officers, whose relations to each other are influenced by the ranks of their husbands. Because Marcia Weidner is the senior wife, the SOA and her husband expect her to attend to the welfare of the officers' families and to further the transmission of U.S. values to them. To do so, she opens her spacious home to the wives for periodic brunches and teas, and she assists the women in numerous ways, such as driving them around town, accompanying them to church, and providing copies of her booklet *"Qué Buscas"* (What are You Looking For) to those needing prac-tical advice about life in Columbus.

At one luncheon gathering that I attended, eight women assembled in Weidner's living room, which was shaded by large trees on the edge of Fort Benning's golf course. The women brought pieces of a friendship quilt that they were making together and that would remain with Marcia

Weidner after they returned to Latin America. At this meeting, I asked them to discuss their impressions of Columbus with me. Some of the most common responses were that "people respect the law" (*la gente anda dentro de la ley*), "everything works here" (*todo funciona aquí*), and "there is a lot of order" (*hay mucho orden*) in the city, although a few commented disdainfully about the perceived rudeness of African Americans. Coming from women whose husbands had long been assigned the task of upholding law and order in Latin America, these remarks were hardly surprising. Yet they pointed to anxieties about the inconveniences and vulnerabilities that these women experienced in their home countries.

Security was a particularly important issue for Colombian Ximena Castaño,[6] who was very grateful for the tranquility of suburban Columbus and for the time that her family could spend together. She explained that she and her children were forced to live alone for long periods in Colombia because her husband's duties as an intelligence officer took him all over the war-torn country, but, in Columbus, he came home when his class finished in the afternoon. Tears came to her eyes as she described the many times that he missed important events in his children's lives, such as birthday parties and Christmas celebrations. Bitterness spiked her disappointment. Barely concealing her anger, she delineated the difficulties that the family encountered when her husband returned to Bogotá, the Colombian capital, after a prolonged absence and reasserted his rights as the household head. The children no longer viewed him as the primary authority figure, and Ximena's newfound independence, which she had never desired but developed out of necessity, discomfited him. In her spouse's absence, Ximena made her own decisions about what, when, and how to do things, but her husband expected a return to old patterns when he set foot in the house. What he found, however, was the resentment and opposition of his children and wife.

Moving to Georgia and recuperating a more traditional, patriarchal family lifestyle did not resolve all of the family's problems, but the suburbs afforded a measure of security that the family had not known in Colombia. At least, Ximena said, she did not have to worry about her husband's safety, and the family was together. Although other countries did not experience the level of violence present in Colombia, the long family separations described by Ximena Castaño were common events in many militaries. Deployments and continuous reassignments kept families on the move, frequently separated, and often rife with tensions, and they belied the notion of the cohesive "military family" so beloved by U.S. Army propagandists. Variations of Castaño's story were repeated by other women in the group, and several emphasized that for them the impor- tance of the SOA lay in the possibility to live together with their families in a country where the water supply was trustworthy, pedestrians crossed the streets at marked intersections, and a sense of orderliness prevailed.

At the School of the Americas, the CGS wives could also promote the image of the particular militaries represented by their husbands and soften the image of these institutions, the men, and the SOA by making them seem less brutal. As the luncheon in the commandant's residence drew to a close, a Guatemalan woman—the wife of a guest instructor who was teaching leadership to the CGS students—presented me with a souvenir of Guatemala. The souvenir was a piece of advertising put out by the Guatemalan tourist agency INGUAT that featured a picture of a young indigenous Mayan girl on a small card. Tied around the card was a woven bracelet—the kind of degraded indigenous art sold to tourists by im- poverished Guatemalans—and under the picture were the words "We Guatemalans invite you to wear a good will bracelet, as a token of our friendship." None of the wives found anything ironic about the gift nor did the gesture strike anyone as inappropriate or even particularly note- worthy. The Guatemalan woman had in fact brought many of these tokens of appreciation with her to the United States to distribute to ac- quaintances. There was nevertheless a cruel irony in her behavior. The Guatemalan military had murdered tens of thousands of indigenous peo- ple in its thirty-five-year civil war, and the traumatized survivors of the army's scorched-earth tactics harbored deep fears of the military.[7] Yet in the intensely ideological world of the SOA, where history was routinely disappeared, and the boundaries between truth and lies, war and peace, genocide and friendship were constantly under assault, there was in fact nothing special about her gesture.

While wives socialized, their husbands spent a great deal of time with each other in the classroom. They started the CGS course in February 2000 and spent the rest of the year—Monday through Friday, 6:30 a.m. to 5:00 p.m.—listening to lectures, participating in small-group projects, and de- bating aspects of U.S. doctrine. It was in this context, amid the intense personal interaction demanded of them, that U.S. and Latin American students assessed each other and devised ways of working together, or at least tolerating each other. The CGS course coordinator took great care to mix individuals from different countries in an assigned seating arrangement to avoid the emergence of nationalist cliques. Name plaques marked the desks of every individual, and when the class broke out into small, working groups to focus on particular issues, the groups were always nationally heterogeneous to the extent possible.

A great deal of competition shaped the emerging relationships of the CGS students, and nationalism provided the language in which much of this competition was expressed. According to the commandant, "everyone wanted to be number one" at the end of the year. The nationalisms articulated at the SOA were intensely male ideologies, because students equated a nation's strength with the potency of its male-dominated military. Paradoxically, these nationalisms were fed by daily life in the United States and training at the SOA, the very experiences that incorporated Latin American students into a broader, global society. Although international military training allowed Latin American students to participate, albeit briefly, in the "American Dream," it also gave them a wider perspective on the subordination of their countries and armed forces within an international order dominated by the United States. The wealth of the United States and its consumer-oriented citizenry underscored the poverty of Latin America. The power and global reach of the U.S. military highlighted the parochialism of Latin American armed forces. And the ease and frequency with which U.S. Army personnel circulated around the hemisphere reminded Latin American officers of their more limited opportunities. The Latin Americans' appeals to the "imagined community" of the nation, which they claimed to represent, and the solidarity of a fanciful "us" were attempts to negotiate their incorporation into an imperial military apparatus on terms more favorable to themselves. Nationalist rhetoric represented a bid by Latin American students to claim a measure of symbolic equality with the United States, even as they became subservient U.S. proxies and struggled to distinguish themselves from each other. As Calhoun notes: "The idea of nation is inherently international and works partly by contraposition of different nations to each other. Nationalist rhetoric offers a way of conceptualizing the identity of any one country that presumes the existence of other more or less comparable units. . . . It . . . make[s] the local nation a token of a global type, to construct it as equivalent to other nations" (1997, 93, 94).

Controlling the seating arrangement was not the only way that the SOA managed student nationalism. As we saw in chapter 1, hosting celebrations of national holidays and filling the position of subcommandant with Latin Americans from different countries were ways that SOA officials paid homage to nationalist sensibilities, while simultaneously subordinating them to their own agenda. The SOA further controlled nationalism through the high premium that it placed on cooperation and team work, which were essential to an effective military and especially a far-flung, global military apparatus. A student could only establish his individuality and his nationalist identity within the very narrow limits of permissible behavior laid down by the army and the behavioral imperatives prescribed by rank. The result was considerable pressure to conform, as too much individualism could damage a man's reputation. In the realm of leadership, for example, the SOA erected an elaborate system of evaluation in which students appraised each other and were simultaneously judged by faculty members and the course director on the basis of five criteria: communication (the ability to listen to others and express oneself); motivation (the ability to inspire others to act in a particular way); decision making (the capacity to discern, reason, and utilize resources knowledgeably); cooperation (consideration for others and the ability to promote collective initiatives); and interpersonal relations (the ability to assess, educate, counsel, motivate, and delegate opportunities that develop individual potential). The students evaluated each other at the middle and the end of the course, while the assessments by faculty and the course director came at the end. The combined evaluation counted for 25 percent of a student's final class grade, or five hundred potential points from a total of two thousand, and small variations in students' cumulative scores could make the difference between winning and losing an award. Those who played by the rules of the game emerged victorious, and they brought honor to their national militaries without disrupting the imperial hierarchy of armed forces. A week-long instructional block on psychological operations demonstrated some of the ways that student relationships played out in the classroom setting.

Psyching Out the Enemy

In U.S. military doublespeak, psychological operations, or PSYOP, are all about "advertising." Repeated by the commandant and several PSYOP specialists in the CGS class, this characterization nevertheless misrepresents activities designed to exploit the fears and vulnerabilities of targeted groups. The PYSOP manual describes four broad goals of psychological operations: to support the political systems of the United States and its allies; to attack the legitimacy and credibility of the enemy; to publicize the reforms and programs implemented for the benefit of the population after a military victory; and to win the loyalty or induce the surrender of enemy forces (Secretaría del Ejército 1993). The manual discusses specific, well-known examples of how these goals are to be accomplished, such as the distribution of leaflets, shows of military force, and civic-action programs. Almost as an aside, however, the manual notes that almost anything has a psychological effect and that "only the limitations of the supported unit when it undertakes an action and the imagination of the PSYOP planner restrict the variety of psychological operations" (I-20). In other words, anything goes in psychological warfare.

On the first day of PSYOP instruction, students saw a short video about the 1991 Persian Gulf war. The video described how the "battlefield was prepared for victory" through the use of fliers and radio programs that were devised to convince Iraqi troops that they could not win. As the class discussed the film, a debate arose about the word "propaganda" and its appropriateness for describing a psychological operation. Discussion focused on the Spanish and English meanings of the term, which is spelled the same in both languages. English speakers understood propaganda to be rumors and information spread deliberately to help or harm a particular individual or group, but Spanish speakers conceptualized propaganda as simply publicity or advertising with fewer of the manipulative overtones of the English word. Although the class was conducted entirely in Spanish, Dominican Republic–born U.S. Major Estrella insisted that students use the word "producto," or product, to avoid the confusion and any implication that a PSYOP might not be true.

The instructor then began a slide presentation of general PSYOP techniques, noting that PSYOPS are always directed at foreign audiences and proscribed for domestic use in the United States. Bolivian Major Oscar Pacello requested clarification. The instructor, who was a Latin American, replied that U.S. citizens "are very sensitive" and might feel manipulated for political purposes if the U.S. military used PSYOPS on them. Pacello was not satisfied and persisted. With a segment of lectures and seminars on human rights still fresh in his mind, he expressed concern about the human rights implications of PSYOPS that were based on fear and manipulation. He wanted to know what international laws regulated psychological warfare and added that he had been unable to find any specific material to guide him on the topic. PSYOPS, he said, could be problematic when a powerful group used them against semiliterate peoples. Because of the potential ethical implications and their repercussions on members of the armed forces, Pacello felt that PSYOPS were like an "atomic bomb" waiting to go off. His question stumped the instructor, who referred Pacello to the training manual, and his persistence seemed to irritate some of the students. U.S. Major Torres—the intelligence specialist with extensive Central American experience—jumped into the discussion. "There is no law," he asserted. "U.S. military doctrine guides the behavior of the United States." This remark provided the definitive answer to Pacello's question.

Pacello said nothing, and despite the strong nationalist sentiments of the Latin Americans, none of them voiced objections—at least publicly—about the obvious double standard: the U.S. military assumed the right to intervene in the domestic affairs of their countries to mount psychological operations, but it claimed not to do so against citizens of the United States. Moreover, nobody commented on the arrogance of Torres's comment. These men understood how power operated and where they stood in the international military division of labor. Furthermore, U.S. doctrine was, indeed, presented at the SOA as a source of truth to which all could turn for enlightenment. During a break, I asked some students if they ever found class discussion objectionable. A U.S. Air Force pilot scoffed at PSYOP specialists like Estrella, who insisted that their "products" were truthful. "States just don't operate that way," he insisted. The Latin Americans, however, were more circumspect. A few told me that they were at the SOA to learn U.S. doctrine, not to debate the policy of the United States, but one officer, who I encountered several months later in Latin America, confessed that some of the things he heard made him "want to pull out his hair." He did not elaborate.

The classroom exchange was just one of many lessons about the hemispheric hierarchy of power and the place that the Latin Americans occupied within it. Yet within this context of hierarchy and paternalism, U.S. and Latin American students did find considerable common ground on which to work. Undermining guerrilla insurgencies, for example, was a concern shared by everyone, and students spent much of their time in small groups, evaluating hypothetical insurgent scenarios, identifying potential PSYOP targets, and situating their operations within larger political and military contexts. On one occasion, the instructor presented small groups of students with a hypothetical situation: Choloma, a village in an unnamed country, had come under the control of a guerrilla movement from "New Palestine," which received external support from the country of "Kracovia." The guerrillas had recruited several informants who supplied them with information about the activities of the local government, and they had won the support of many people through friendly face-to-face relationships and assistance in resolving village problems. They had also murdered a town councilman in an attempt to further consolidate their control of the village. The exercise provided additional information about the relationships between the villagers and the military, local landowners, and the town council, and then asked students a series of questions that, as PSYOP specialists, they would have to address in a real-life situation. What group or groups would be the best target of a psychological operation, and what were their vulnerabilities? What would be the most important objective of such an operation? What were the desired outcomes of the operation?

The banality of student discussions tended to obscure the fear and disorientation that an actual psychological operation could generate among targeted people. For example, the men in one group—a Dominican, a Bolivian, a Venezuelan, and three from the United States—spent an enormous amount of time mired in a discussion about how to distinguish between the national-level political objectives and the concrete goal of the PSYOP mission. What, for example, were the duties and responsibilities of the PYSOP officer? Where did they end, and where did the duties of other authorities begin? Answering these questions, they felt, was important in order to properly fill out the requisite army paperwork for a psychological operation to begin, and a U.S. officer with extensive PSYOP experience in Kosovo held forth in broken Spanish about the appropriate distinction, slowing the group's progress considerably. Their consultations tended to naturalize and trivialize the deadly serious nature of war in general and the violent, manipulative tactics in which PSYOP specialists like themselves engage. These officers needed to locate themselves within the hierarchy of power and the chain of command. Then, they could turn their attention to the dispassionate task of bureaucratic problem solving and formulate an efficient solution that responded to the dictates of a higher authority. In this way, students relieved themselves of sole responsibility for their actions.

Psychological Operations was just one of several instructional blocks in which the CGS students participated before they graduated at the end of the year and scattered across the hemisphere. A few of the Latin Americans would remain at the SOA for an additional year as guest instructors, but most returned to their respective countries and, in many cases, were quickly promoted. Colombian Major Espitia returned to work with a militarized police force in the city of Cali. U.S. Major Torres began an unspecified job in intelligence at the Pentagon. Another U.S. graduate took up a teaching assignment at Fort Benning. Many other U.S. students were posted around Latin America in a variety of assignments that corresponded with their particular branch designations. For the latter, U.S. embassies and their associated Military Groups were common destinations, but in some cases of counterintuitive bureaucratic logic, the military shipped graduates to other parts of the world. That was what happened to Lieutenant Colonel George Ruff, head of the SOA's Department of Joint and Combined Operations, who graduated in the early 1990s and went to Korea before receiving an assignment in Latin America.

Neither the U.S. nor the Latin American officers remained in any position for a long time. After two or three years, they were reassigned and on the move again, and, as they circulated from job to job around the hemisphere, former classmates reconnected with each other and CGS graduates from different cohorts. Many found themselves once again working together. U.S. officials liked to think that the Latin American graduates had a better understanding of "how the United States does business." The graduates were also known to many U.S. Army officials, who viewed them as important liaisons to the various militaries of Latin America. Their ascent through the ranks was generally welcomed by U.S. personnel because it represented—at least in theory—the presence of individuals in the senior ranks of the armed forces who were willing to work with the United States military and understood its doctrine. The case of Oscar Pacello, who graduated in 2000, and Rand Rodríguez, a 1996 graduate, provide an example.

Doing Business the American Way

Oscar Pacello returned to Bolivia after his graduation in late 2000. The army promoted him to lieutenant colonel, and he assumed new duties as the liaison officer between the Bolivian military high command and the

U.S. Military Group based in the embassy. Pacello was the latest in a series of CGS graduates from the School of the Americas to hold the position. The year-long posting with the Military Group was a plum job because, as a U.S. official explained, it allowed Pacello to "work over in the know group," that is, with the military that really called the shots in Bolivia, but it also made Pacello the target of envy and the brunt of jokes among his Bolivian colleagues. Scathing commentaries circulated in a clandestine Bolivian army publication that described how Pacello's lips constantly affixed themselves to the rear ends of his gringo masters. When I met up with him again in Bolivia in 2001, Pacello was in fact unabashed in his admiration of the United States and in his gratitude for the opportunity to experience "a more organized way of life" in Georgia. Without alluding to dissidents' remarks, he criticized those in the Bolivian army who blamed all their problems on "the gringos" and attributed these feelings to the blind nationalism that emerged from the military academy and the leftist discourse of the universities, where, he maintained, "four out of seven classes teach Marxist political theory."

Pacello worked with the U.S. Military Group on the fourth floor of the bunkerlike headquarters of the embassy. Constructed in the mid-1990s, this unsightly structure rose like an ugly beast from the stately Avenida Arce and clashed with the apartment buildings and turn-of-the-century mansions in the surrounding neighborhood. Its construction raised the hackles of well-heeled local residents because the original plans made no room for sidewalks on a street that bordered the emerging monstrosity. Although residents won the fight for sidewalks, they had to resign themselves to living next to the architectural equivalent of an eight-hundred-pound gorilla. Encircled by high walls of reinforced concrete, the embassy bristled with antennas protruding from the roof, and narrow slits covered with reinforced glass served as the only windows. A heavily fortified security booth screened visitors and employees. All of this spoke volumes about the real and imagined antipathy that the United States provoked among local people.

Pacello's primary contact was fellow SOA alumnus and CGS graduate U.S. Lieutenant Colonel Rand Rodríguez. Rodríguez was serving out the last six months of a two-year assignment in Bolivia when Pacello returned to La Paz, and he was preparing to move to a new posting in the United States. He had learned a lot about the Bolivian military in particular and about Latin American forces in general from his travels around the region.

"So many things are institutionalized in the Latin American armies," he explained, "and that goes from paternalism, corruption, to the [list] goes on. . . . Their armies are based more on friendship [than ours]. The positive aspect is that you get to know personalities, but the negative is that it influences the way they do business. If I need something in the United States, it's based on rank and the way I work. . . . But in Latin America, I'm a lieutenant colonel, he's a lieutenant colonel, he's my friend and he'll help me get done what I need to get done, [because he is my friend] not because it's the right thing to do. You know what I'm saying?" The cgs course gave Rodríguez an initial entrée into this world of friendship and patronage that served him well by the time he reached Bolivia. He elaborated:

> I went to the School of the Americas and met a number of people, and many people from the Bolivian army. There was a Bolivian police officer and an army colonel. The guy I sponsored was an officer from El Salvador, and I also sponsored a lieutenant colonel from Mexico. . . . From my point of view, that's great. [It's] where you make your connections as a foreign area officer. You go in, you talk to people. They're your classmates. You learn about what they do and what it is like down in specific countries. . . . When I came to Bolivia, my friends came to see me and helped me do the introductions. They break you past those initial barriers that you would normally find, especially in Latin American countries. You know—"he's an American, he's a gringo. But I don't want to be seen as an American lover." So they'll come and say, "I knew him at the School of the Americas. I spent a year with him. He's a great guy. I know his family." That breaks down the barriers.

Having Pacello to "break down the barriers" with Bolivian commanders was important to U.S. military personnel in Bolivia who had to contend with a country in a continuous state of turmoil. Peasant road blockades and demonstrations against privatization of the public water utility paralyzed the country for three weeks in September 2000, and during the subsequent months the level of discontentment had scarcely abated. This was because fifteen years of radical, free-market economic policies had taken a heavy toll on peasants and working people, and the aggressive, U.S.-sponsored coca eradication campaign in the Chapare region had undermined one of the last economic supports available to peasants (see chapter 7). To make matters worse, the corruption and heavy-handedness of the second Banzer administration had alienated many people.

Faced with the ongoing turmoil, U.S. and Bolivian military officials wanted to maintain the capabilities of several highly trained Ranger and Special Forces detachments that operated from strategic locations around the country. These elite units provided intelligence, located coca fields and cocaine laboratories, cleared road blockades, and "responded to any emergencies within the country." In Rodríguez's opinion, however, several of them "still needed a lot of work," a view that was shared by high-ranking Bolivian commanders who had seen Special Forces demonstrations in the United States and who wanted to increase the capabilities of their units in Bolivia. When Bolivian commanders prepared to send their first soldier to the U.S. Army's Special Forces School at Fort Bragg, North Carolina, however, a problem arose for Rodríguez.

The Bolivians passed over the individual preferred by Rodríguez and put forward another person, a soldier related to a powerful official who pulled strings to get his candidate to the United States. Rodríguez explained:

> We found a guy who we thought had all the capabilities—good physical condition and command of English—to do the course. We were waiting for his name to come up through the official channels, and then one day, I walk into work and find a Bolivian officer there. I say, what are you doing here? He says that he's testing for the Special Forces. And I think okay, well, why not. He took the English test and he scored a 40, minimum qualification was 80. A couple of days later it was his name that came in as the official representative of the Bolivian army.

Rodríguez was not happy, and he called the commanding general of the Bolivian armed forces to discuss the matter, but in the end, Pacello proved instrumental in resolving the problem. In a meeting with Bolivian officials, Rodríguez and Pacello made a case for the U.S. choice. This was a sensitive task because the choice of whom to send to a U.S. training program rested with the local armed forces, and Rodríguez did not want to appear to be dictating a decision to the commander. Pacello, however, was a Bolivian and he could be more insistent, as he helped Rodríguez negotiate the fragile sensibilities of rank and nationality. In the end, the two men managed to convince the Bolivian commander to send the U.S.-backed candidate. Rodríguez acknowledged his debt to Pacello. He concluded that "having an SOA graduate who understands what we're talking about and the way we do business is an asset for us. They help us convince [their militaries] of the way we do things."

Doing things the American way entailed the formation of complex, power-laden relationships through which U.S. officials cajoled, manipulated, and made demands of Latin American armies. And, as Rodríguez indicated, having a local officer like Pacello, who either agreed with U.S. policies, or benefited from the rewards that collusion offered, made a great deal of difference in "getting the job done" for the Americans. The SOA, and particularly the CGS course, cultivated working relationships between U.S. and Latin American officers, and it facilitated U.S. efforts to dominate regional militaries with strong nationalistic sensibilities.

As successive cohorts of CGS trainees and graduates from other SOA classes took their military training and circulated around the hemisphere, serious allegations of human rights violations frequently followed in their wake. Before leaving the CGS class of 2000, we must therefore consider how the SOA addressed the issue of human rights with the trainees, and how the latter engaged this instruction. The next chapter explores these concerns.

CHAPTER 6

Human Wrongs and Rights

The human rights abuses of cold war Latin America finally came home to haunt the School of the Americas in the 1990s. The release of a list of some 60,000 SOA graduates in 1993 revealed the names of some of the hemisphere's most notorious dictators, death squad operatives, and assassins, and when human rights activists began comparing these names to those listed in a variety of truth commission reports, the results were startling: SOA graduates took part in some of the worst human rights atrocities of the cold war. For example, ten graduates participated in the massacre of nearly one thousand people in the Salvadoran village of El Mozote, and two others stood accused of the murder of Archbishop Oscar Romero, after he pleaded with soldiers to stop killing their own people.[1] One hundred twenty-four, or nearly half, of the 247 Colombian officers cited in 1992 for gross human rights violations trained at the SOA (OMCT 1992), and men like the Argentine Andrés Francisco Valdéz not only tortured prisoners in unspeakable ways but also tormented their families during truth commission hearings (Feitlowitz 1998, 268–69). These men were just a few of many other SOA graduates who stood accused of terrible crimes. The declassified list of graduates did not provide a direct connection between SOA training and the subsequent human rights violations perpetrated by alumni, nor did it clarify how these men caught the attention of commanding officers early in their careers and secured a spot at the SOA, but the list raised disturbing questions about the SOA's training methods, the people it trained, and, more broadly, U.S. policy in the hemisphere.

Even more than the list of graduates, the exposure of torture manuals in 1996 fueled a budding social movement that demanded the School's closure. The movement placed the formerly obscure institution under the harsh light of public scrutiny and forced SOA officials to defend themselves against accusations that they were training torturers (see chapter 2). Critics drew a direct line between the SOA and the brutal practices of Latin American security forces, and they opened the door for thinking about the United States as a perpetrator of terrorism. Viewing the United States and its military institutions in this way has always been subject to strong resistance, and, when the SOA initiated an "expanded human rights program" in the late 1990s, it was just one example of its refusal to account for past practices and ongoing abuses.

Human rights instruction at the SOA is less about curbing the atrocities committed by security forces than shoring up the legitimacy of a discredited institution and obscuring the brutality of U.S. foreign policy. More broadly, however, the SOA human rights program is part of an ongoing struggle over how the cold war past is remembered, whether the United States is perceived as an empire and a perpetrator of terrorism, and how much impunity the United States military apparatus and its Latin American allies can claim in the present. Cold war memories are not only shaped by what happened in the past. They also emerge from contemporary social struggles that mold how historical events are recalled, forgotten, distorted, invented, and reworked.[2] What is remembered and forgotten is a question of power, and impunity—the ability to operate without fear of punishment—is an aspect of power that must be continually defended, regenerated, enforced, and maintained to protect current practices.

Officials at the School and in the Defense Department faced several challenges as they contemplated a human rights program for the SOA. They had to placate an array of critics or face the possible closure of the institution, but they also confronted the ire of Latin American soldiers who bristled at the notion of U.S. officials lecturing them on human rights. A serious examination of human rights violations in Latin America should in fact raise questions about the United States' own involvement in training allied militaries, but the SOA downplayed the connections between the United States military and the crimes committed by security forces trained by it. Officials dismissed the atrocities committed by graduates as either the actions of "a few bad apples," who never understood what the SOA taught them, or a *natural* inclination toward brutishness found among Latin American security forces. They adopted a paternalistic approach to human rights training that targeted the Latin Americans but carefully avoided any direct discussion of the dirty wars that ravaged Latin America in the twentieth century and U.S. involvement in them. Not surprisingly, the SOA's effort to inculcate human rights never examined U.S. policies and justified the perpetuation of impunity, while simultaneously legitimizing the continued training of Latin American soldiers and the militarization of the hemisphere.

The SOA human rights program is part of an argument about the past, present, and future of the Americas that is best understood through the lens of the Command and General Staff Officers Course (CGS). The week-long human rights section that unfolded in February 2000 displayed the SOA approach to training ethical, "professional" soldiers, a key justification of the School's current mission. It also illustrated how the SOA silences and effaces cold war human rights violations by skirting analysis of the Latin American dirty wars and avoiding key issues of justice and accountability in a showcased session on the 1968 My Lai massacre of Vietnamese civilians by U.S. soldiers. The course did little to challenge entrenched student beliefs about human rights and their defenders, as it prepared them to return to countries where human rights were routinely violated, nor did it challenge the pervasive impunity enjoyed by the armed forces. Yet by appearing to engage the issue of human rights seriously, while ignoring or marginalizing the history of dirty wars, the SOA attempted to justify its mission to a growing number of critics.

Human Rights and the Professional Soldier

SOA officials insist that a concern for human rights is embedded in U.S. military doctrine and that this concern has always been communicated to Latin American students through a variety of courses. Nonetheless, with the creation of its human rights program in the mid-1990s, all SOA trainees now receive from eight to forty hours of instruction focused specifically on human rights. The amount of time depends on the length of the course: short courses (four weeks or less) receive an eight-hour instructional block; medium-term courses (four weeks to six months) get twelve hours, and the longest courses (six months to one year) require forty hours.

In early February 2000, SOA officials busied themselves with preparations for Human Rights Week, the first and flashiest segment of the Command and General Staff Officer course. The week's activities fell into three parts: the first covered ethics and army values; the second examined international humanitarian law and its applicability during military operations; and the third covered military operations and the issue of legal and illegal orders, ending with a discussion of the 1968 My Lai massacre in Vietnam. Officials wanted the events of Human Rights Weeks to accomplish a number of goals, but the most important objective was good public relations. They needed to convince a skeptical Congress and public that the soa was not averse to examining human rights issues and that it took very seriously the human rights training of its Latin American students. To this end, the School issued various press releases about the scheduled sessions and invited local television and newspaper reporters to attend.

Commandant Weidner kicked off the week's activities with a presentation titled "Military Ethics and the Use of Force" that emphasized the notion of the "professional soldier" and "just war doctrine." His talk was scheduled for 8:30 a.m. Students awaited his arrival in a large classroom, where they sipped coffee and milled around with other officers from a training course for human rights instructors, who would attend the sessions with them. Rows of long desks and comfortable swivel chairs filled the room. When the commandant entered the room and approached the podium to begin his talk, the men took their assigned seats, and as Weidner adjusted the microphone and cleared his throat, the room quickly quieted down. What followed was an ideological exegesis on the justifications and rationales for the existence of the military, one that made assertions about the differential qualifications of certain peoples and defined a special role for the armed forces that set them apart from other groups.

Speaking in flawless Spanish, Weidner proposed a view of societal ethics that emerged from "customary norms." These norms, which approximate what anthropologists used to call "culture," that is, an interwoven system of meanings shared by a people, represented less the imperatives of a dominant class than the collective consciousness of society writ large. They formed a corpus of timeless, immutable traditions that were communicated to individuals through the institutions of family, church, and school but changed little, if at all, in the process of transmission. Although everyone shared customary norms in Weidner's formulation, threats to the maintenance of these conventions constantly arose from within society. Traditions therefore required continual reinforcement

through the establishment of laws, the application of social pressure, and
the threat or the actual use of force, and the armed forces assumed a
crucial role in compelling people to behave properly. Weidner thus presented a view of Latin American militaries that assigned them the task of
enforcing the domestic status quo rather than defending the nation-state
against external aggressors. This perspective had been commonplace in
the United States military since the cold war. Weidner also made little
room for other perspectives and practices that arose from within the
"customs" of class-divided, Latin American societies to challenge the received wisdom of dominant groups. His speech resonated with a long
tradition of conservative political thought that celebrates social hierarchies and extends from Edmund Burke to Jeanne Kirkpatrick and Samuel Huntington.

The commandant then moved on to a consideration of the "professional soldier," a figure, he asserted, that is central to a modern military.
The notion of the professional soldier is also key to the image of the SOA
that Weidner and other officials like to project. Professionalizing Latin
American soldiers, they say, is one of their most important missions.
Officials do not deny that militaries—and even SOA graduates—perpetrated human rights violations in the past. Yet they attribute these trans-
gressions to the unprofessional behavior of a few individuals who be-
smirched the honor of their militaries, and they argue that creating better
professionals, and especially a competent officer corps, is one way to
prevent these excesses in the future. The argument places the burden of
criminal responsibility for human rights crimes on Latin Americans and
obliterates any consideration of the close relationship between Latin
American and U.S. security forces. It also presents the notion of military
professionalization as something new and important in the wake of widespread dirty-war abuses, even though the School of the Americas has
trained military "professionals"—career officers—since its inception in
Panama.

Drawing heavily on the work of Samuel Huntington, one of his favor-
ite guides on intellectual matters, Weidner defined the professional soldier as less a cool, efficient killer than an expert in the "management of
violence," and he assigned a chapter from Huntington's book *The Soldier
and the State* to the class as required reading. Huntington, who is best
known for his pronouncements about a "clash of civilizations" between a
civilized West and a barbarous East, distinguished professional military

managers from the mass of enlisted personnel in an important way. While the former managed violence, the latter "applied" violence but had "neither the intellectual capacity nor the professional responsibility of the officer."[3] For Huntington and Weidner, the professional soldier was thus similar to doctors, lawyers, businesspeople, and other highly trained individuals distinguished by their capacity for intellectual and managerial work. He made the ultimate decisions that would determine whether the intentional killing of humans was permissible, and he determined the answers to key questions: Was the cause defensible? Were nonviolent alternatives exhausted? Was there a reasonable expectation of success? And was the planned response proportionate to the threat? If the answers to these questions were affirmative, then the war was considered just.

Yet it is doubtful that a war can ever be truly just when it involves the killing of civilians, when it displaces hundreds—even thousands—of people, and when it sustains a system of gross inequality. This is particularly the case in contemporary Colombia, which has sent more students to the SOA than any other Latin American country. Colombian military and paramilitary forces have murdered and disappeared tens of thousands of civilians whom they view as dangerous and undeserving of basic rights (see chapter 7). This is state terrorism, not a just war, and state terror—violence aimed at citizens by their own governments—has long been the central problem in Colombia and elsewhere in Latin America. It has never been waged on behalf of a just cause. Hobsbawm notes that governments rarely go to war because they deem it just to do so. They go to war over concerns for national interests, and they claim that their actions are just to win public support (2000, 16–17).

Focused armed resistance against an immediate attacker is tenable under certain circumstances, but war is indiscriminate. It always leads to death through the direct targeting of civilians or through "accidents" and "collateral damage" when loosely defined "military targets" are attacked.[4] And warfare generates widespread suffering and additional casualties through the destruction of infrastructure and the environment, which leads to public health problems and economic chaos that cannot be described as coincidental.

The just-war thesis is a tactical anodyne that obscures the horrors of war through its emphasis on the responsible use of violence by military professionals.[5] Conceptualizing themselves as high-minded professionals who fight just wars undoubtedly soothed the anxieties of many men in that room, listening to Commandant Weidner. Images of the professional soldier engaged in the good fight enhanced the virility and heroism of men who sacrifice their personal safety to fight a malicious enemy for the greater benefit of the nation.[6] They also permitted officers to distance themselves from their humble social origins and to dissociate from the real terror of actual violence, which was passed off to front-line soldiers recruited from racial and ethnic minorities of supposedly inferior intellect. In this way, officers could claim membership in a genteel middle class, while they continued to control the machinery of coercion.

By the time Weidner finished his moralizing sermon, he had exalted his listeners and reaffirmed their importance. He had also spelled out the military's mission in a supposedly democratic society by conjuring up an imaginary world governed by the benign force of custom, to which all citizens freely subscribe, and then asserted that such a world actually existed. He proceeded to assign the armed forces the task of upholding customary norms and social ethics through the professional management of violence, and he further maintained that the military discharged this responsibility on behalf of society itself. Although such ideological assertions would certainly have invited challenges elsewhere, they served as the foundational premises to which the events of Human Rights Week ultimately referred.

Human Rights Week then moved ahead with little, if any, discussion of specific Latin American examples, nor did instructors use illustrations from Latin America to make points. When lecturers descended from the realm of metaphysics, they typically drew on hypothetical situations or the United States' experiences of conventional war in Europe or the Middle East for specific examples of general principles. This was an approach that typified much of the SOA's pedagogy, and it was justified by the commandant, who insisted that Latin Americans came to the SOA to associate with the world's most powerful military. He claimed that they were not always interested in the experiences of other Latin American nations, which, in some cases, they viewed as inferior to their own, and putting too much emphasis on one country risked offending the nationalistic sensibilities of those whose homelands were ignored. These considerations were undoubtedly important for men like the commandant whose main objective was to steep trainees in U.S. military doctrine and build hemispheric unity. Yet in the context of Human Rights Week, the avoidance of Latin American cases removed the spotlight from specific militaries, their policies, and the relations between them and the United States at particular times.[7]

A session on international human rights and humanitarian law led by a Red Cross representative who resided in Guatemala was fairly typical in this regard. The instructor initiated a discussion of the rules of war, the difference between combatants and noncombatants, and the treatment of prisoners of war, and he situated his remarks within the guidelines established by the Geneva Convention and its associated protocols established after World War II. Students then broke up into small groups. They received handouts that presented a variety of problems raised by hypothetical combat scenarios, and each group had to come up with practical solutions that did not violate the Geneva Convention. One exercise, for example, presented the case of an imaginary town besieged by guerrillas. It required the students to make decisions about how to protect the town from attack and deal with civilian collaborators. "We're going to kill them with human rights," snickered one indifferent individual. In another exercise, the men divided into two warring armies and had to make a number of decisions within a context of escalating combat. If, among their prisoners, they encountered an armed civilian who had participated in combat, how should the individual be treated? If one of their soldiers killed an enemy combatant in cold blood after the individual had thrown down his weapon and raised his arms, what was the proper response? How should an officer from a third nation, captured in combat, be treated if he defined himself as a "technical advisor"? Students answered these questions by citing articles from the Geneva Convention, and in this way the session proceeded much like a law school course.

A subsequent class discussion about torture developed in similar fashion. The Red Cross representative made it clear that torture was both immoral and a violation of the Geneva Convention, but the main thrust of his argument was that torture did not work—an approach, I later learned, that he based on the assumption that torture was widespread among Latin American security forces. He pointed out that prisoners say almost anything under torture to stop their torment, and the result was a lot of false information that wasted the time of those sent to verify it. All of the students did not concur. One argued that in some cases torture could, indeed, be an effective means of gathering information. Another asserted that torture was not only effective but necessary in certain circumstances. He described the so-called ticking bomb scenario in which people would die unless a prisoner divulged knowledge about its location. The instructor dismantled each allegedly hypothetical argument. International and U.S. law, he said, did not contain any exceptions condoning the use of torture, even in cases where life and death appeared to hang on the information held by detainees. Moreover, the "bomb scenario" rarely occurred in practice. Permitting torture under such circumstances, the instructor asserted, provided no limits on how much or what kind of torture was allowable, and exceptions could easily become standard practices. He did not convince everyone.

Despite the importance of these issues, the session—as well as others like it—skirted the ugly details of particular Latin American cases. Its emphasis on preparing students to respond appropriately to particular combat situations effaced history by ignoring how and why specific individuals and their armed forces violated the human rights of others—especially unarmed combatants—at particular times and in distinct places. With the exception of Colombia, Latin American militaries have generally ended, at least for the present, the practices of disappearance, massacre, and extrajudicial executions that characterized their cold war pasts, but they have been extremely reticent about allowing themselves to be held accountable for their actions. Commanders typically claim that the past is best forgotten. They assert that revisiting conflicted periods of national trauma, when their countries were at war, simply opens old wounds and undermines efforts to construct orderly, democratic societies, and they use evasions and threats to dissuade civilians from pressing human rights issues. Not surprisingly, widespread impunity is the norm throughout Latin America. Although truth commissions in some countries have made limited progress in clarifying the fate of the murdered and disappeared, they have been generally unsuccessful in holding military perpetrators accountable (e.g., Hayner 2001). But without an accounting and a memory of particular events, it is difficult to understand their legacy, and human rights training at the soa did little to examine or clarify the past.

Silencing the past did not stop with the avoidance of Latin American dirty wars. It also extended to the erasure of key aspects of U.S. imperial history beyond Latin America. Sensitive to charges of hypocrisy and posturing from Latin American students, and demands by human rights organizations that the U.S. military lead by example, soa officials organized a panel discussion of the My Lai massacre to cap the events of

Human Rights Week. According to the commandant, the panel would serve a number of purposes. It would demonstrate to the Latin Americans that the United States could examine its own mistakes and hold itself accountable. In addition, the extensive postmassacre investigations offered rich materials to explore how illegal orders were given and carried out in the context of an unfolding military operation. SOA officials also hoped that the session would garner favorable media attention in the ongoing propaganda campaign against critics.

The panel convened in Pratt Hall, a spacious theater with room for approximately two hundred people. As students filed into the auditorium at the appointed hour, it was apparent that the Public Affairs Department had scored a minor victory. A cameraman from the Fox television news station positioned his camera in the back of the room, and at least one representative from the print media was in attendance. Charlie Liteky, a former priest and Congressional Medal of Honor recipient, slipped into a front row seat. Banned from Fort Benning for his anti-SOA protests, Liteky had received special permission to attend the event from the base commander, who perhaps sensed a public relations payoff, and he remained under the watchful eye of a military escort.

Two panelists would speak about the My Lai massacre: former helicopter pilot Hugh Thompson and his gunner, Larry Colburn. Thompson and Colburn had flown over My Lai on March 16, 1968, as the massacre was in progress. They saw bodies strewn across the ground and witnessed U.S. soldiers repeatedly shooting civilians at point-blank range. An enraged Thompson landed the helicopter between one group of soldiers and Vietnamese civilians, who were mostly women, children, and elderly men, and he ordered Colburn to fire on the infantrymen if they resumed the killing. The troops backed down, although the bloody rampage continued in other parts of the region.

For years after his courageous action, Thompson was reviled in the United States. Bumper stickers demanded his court-martial, and a congressman demanded that he go to jail for threatening the lives of American servicemen. House Armed Services Subcommittee Chair L. Mendel Rivers even asserted that no massacre happened in My Lai and that claims to the contrary were efforts to discredit the Vietnam War. Thompson remained an unsung hero for some thirty years. His public image only improved in the late 1990s with the publication of a sympathetic book about his life and a story by the CBS news program *60 Minutes* that covered his return to My Lai on the thirtieth anniversary of the massacre. The army finally awarded him the Soldier's Medal—the highest award for bravery not involving the enemy—in 1998. Although the official recognition was a long time coming, the irony associated with the Soldier's Medal was inescapable, given the identity of "the enemy" in My Lai in 1968.

Thompson showed the *60 Minutes* film clip to the CGS class before he began to speak. He appeared nervous. Agreeing to participate in the SOA event had not been an easy decision for him. Speaking in a deep Southern drawl, he began his remarks with a careful disclaimer that he neither supported nor condemned the School.[8] He then launched into a moving description of what took place in My Lai on March 16, 1968. He described how Charlie Company, which came to Vietnam in 1967 under the command of Captain Ernest Medina, was assigned the task of pressuring the Viet Cong in an area of Quang Ngai Province dubbed "Pinkville" by the Americans. On March 14, a booby trap surprised a squad from Charlie Company and killed several soldiers, while blinding another. The deaths outraged everyone, and, according to Thompson, the soldiers went into My Lai two days later for revenge. By the end of the day on March 16, some five hundred Vietnamese were dead, massacred in cold blood by two platoons under the commands of Lieutenant William Calley and Lieutenant Stephan Brooks. But the soldiers, Thompson insisted, did not stop with murder. They also raped women, shot livestock, poisoned wells, and burned houses. He emphasized that those who perpetrated the massacre knew that what they were doing was wrong, but many, he believed, had succumbed to peer pressure. Yet he quickly pointed out that everyone did not take part in the mass murder. One soldier shot himself in the foot to get evacuated from the area, and other soldiers avoided My Lai because they knew what was happening.

Thompson told the group that the commanding officers at My Lai were bad leaders who gave illegal orders to kill unarmed civilians. Twenty-four-year-old William Calley idolized his immediate superior, Captain Ernest Medina, and constantly sought his approval. And Lieutenant Brooks, Thompson asserted, "made Calley and Medina look like good guys." When Thompson finished his presentation, he had clearly moved the students, who gave him a standing ovation. Yet the ensuing discussion never clarified how soldiers faced with commanders like Calley, Medina, and Brooks could deal effectively with illegal orders; indeed, if one soldier had been moved to self-mutilation to avoid complicity, a direct refusal to obey orders was probably out of the question and perhaps even life threat-ening. More importantly, if in fact it was U.S. policy to kill civilians in areas under presumed enemy control, that policy was a crime against humanity in the clearest sense. But no one suggested a broader examination of U.S. policy in Vietnam, nor did anyone address the issue of justice.

Much like the 1981 Salvadoran army massacre of civilians in El Mozote, or the Guatemalan army's numerous massacres of indigenous Maya in the western highlands during the late 1970s and 1980s, the U.S. Army perpetrators of mass murder at My Lai were not held accountable for their crimes. During the weeks and months after March 16, the army tried to cover up the event, but as news of the killings became public, pressure for an open, independent investigation grew in the context of a strong antiwar movement. The Nixon administration opted for a closed-door Pentagon inquiry and named three-star General William Peers to head an investigatory committee. The committee's findings—known as the Peers Report—concluded with a recommendation for action against dozens of men who participated in rape, murder, and the subsequent coverup. But in the end, only a handful of people were tried and Lieutenant Calley was the only person convicted. Many infantrymen were immune from pros-ecution, because they were out of the army by the time the controversy erupted and no longer subject to the Uniform Code of Military Justice.[9] High-ranking career officers remained untouchable behind a wall of si-lence, coverup, and lies.

Calley, too, would eventually benefit from widespread military im-punity. Court-martialed at Fort Benning and found guilty of the murder of at least twenty-two Vietnamese civilians, he received a life sentence at hard labor. But two weeks after the verdict, President Nixon ordered him released from prison and placed under house arrest. The sentence was subsequently reduced, and the Secretary of the Army paroled Calley in 1975 after he had served less than three and a half years of house arrest. Thirty years after the massacre, while the SOA panel revisited his actions at My Lai, William Calley lived quietly in Columbus, across town from the SOA. Balding and heavier than in the early 1970s, he worked in his father-in-law's jewelry store and occasionally visited Fort Benning to sell rings to the graduates of a parachute class.

Engaging the My Lai massacre was not particularly threatening for the SOA. Unlike the Latin American dirty wars, which are still controversial, the silencing and obfuscation of U.S. Army atrocities in Vietnam was much more complete. This silencing took many forms, including seg-ments of the American public that concerned themselves more with U.S. casualties than Vietnamese victims, the absence of a powerful domestic human rights movement to demand justice and accountability, and the intense right-wing nationalism of the 1970s and 1980s. It reached an ugly nadir in the spring of 2001—after the CGS students had gone home—when the *New York Times* reported that New School University president and former senator Bob Kerry had led a Navy SEAL team that massacred at least thirteen Vietnamese civilians in 1969. The ensuing public debate focused less on the anguish of the victims than the psychological torment of Kerry, who became an object of pity.

Although the panel opened the door for a discussion of impunity and accountability, the SOA and its assembled students never walked through it, nor did they draw comparisons between My Lai and similar cases from Latin America. The institution was, after all, pushed into adopting a human rights program by a social movement that threatened its very survival, and SOA officials hoped to control the controversy that envel-oped them and to shape human rights discourse on their terms. "Dis-course," writes Eric Wolf, "has its reasons; it also has consequences" (1999, 283). The activities of Human Rights Week silenced and distorted the past, justified the training of armed forces that remain largely unre-formed, and rationalized the continued militarization of the Americas. The most important message was not respect for human rights but the high value of good public relations. This became apparent five months later, when the class traveled to Washington, D.C.

Touring Democracy in the Heart of Empire

It was a typical hot, humid, midsummer night in Washington when the Latin American students of the CGS class arrived on a commercial flight from Atlanta. Ten captains from the Combined Arms Advanced course accompanied them. The men came for one week to tour the U.S. capital's monumental core, to meet with government officials, and to view the apparatus of U.S. political and military power in operation. SOA officials spun the trip as a lesson in democratic governance. The students were eager to see Washington, and some had friends and relatives in the metro-politan area whom they planned to contact. Everyone was relieved to have a respite, if only for a week, from their studies and to have a chance to enjoy themselves. The cautious sizing-up and polite correctness that char-acterized their interactions in February were gone, and the transnational male bonding that was so central to the CGS course was clearly underway. Friendly backslaps, occasional one-upmanship, and casual vulgarities pep-pered the men's interactions. Some cracked jokes. Others talked about upcoming soccer matches between rival Latin American teams. The U.S. members of the CGS course, however, were unable to enjoy the sights of Washington and the respite from the daily grind of class. In the army's estimation, these men understood enough about democracy—indeed, they were symbolic of it. There was no need to waste training dollars on a trip to a city that they already knew, so the U.S. trainees remained in Georgia.

An exhausting list of activities exposed the Latin Americans to a range of political and military institutions. The seven-day agenda included visits to key government institutions—the Congress, White House, Supreme Court, and the headquarters of the Organization of American States—meetings with military officials in the Pentagon, the Interamerican De-fense Board, the Center for Hemispheric Defense Studies, and the Inter-american Defense College, and lessons on U.S. history through visits to the Smithsonian's Air and Space Museum and six war memorials. Al-though the tour concentrated on visits to military institutions and memo-rials, much of the agenda differed little from the typical round of sight-seeing that occupied thousands of civilians who visited the capital every year. On one day, for example, SOA students waited on a long line of hot, sweaty tourists to walk through the White House, and they gawked and snapped pictures in the Capitol rotunda with throngs of other sightseers.

Some activities that were hyped by the SOA public affairs department turned out to be less glamorous than suggested by the hyperbole. A "state luncheon" in the Rayburn congressional building was a rather lackluster affair in which the group was crowded into a cramped, nondescript room and served a mediocre lunch proclaimed by several students to lack flavor. Two Georgia congressmen—Collins and Bishop—eventually made an appearance and offered a few boilerplate remarks extolling the virtues of the SOA. They then shook everyone's hand and left. The SOA entourage later visited Bishop's office, but the congressman, according to a staff member, was on the telephone and unable to attend the students. His young assistant escorted them through the staff quarters, and another aide handed out small bags of Georgia peanuts and cans of Coca-Cola to the group. It was clear that the "future leaders of Latin America" would need time to register on the Washington radar screen as more than glorified tourists.

Students were, however, impressed with the city and the power symbolized by its monumental core. After visiting the Capitol rotunda, Major Pacello marveled at the majesty of the "Anglo-Saxon architecture." Another student felt that Washington exemplified the "leadership qualities" of the United States. Experiencing vicariously the power and the glory of the United States was more difficult for a black Dominican student, who accompanied his classmates to the food courts of the Old Post Office for lunch one day. As the students entered the building, one of them spied a photographer who would take their pictures with either Michael Jordan or Bill Clinton for five dollars. The photographer used actual pictures of Jordan and Clinton and substituted his subjects' heads for those of others shown shaking hands with the celebrities. The Dominican and many of his peers were eager to have such a memento to take home with them, and most chose to be photographed with Clinton. The young man posed proudly for the photographer and then waited eagerly while the picture was developed. When he was handed the photograph, however, his face fell, and a look of disappointment spread across it. "They whitened my hand," he mumbled. Indeed, the hand that grasped President Clinton's in a vigorous handshake was white, and it clearly did not belong to this unhappy soldier who had just parted with five dollars. In many ways, the photograph served as a metaphor for all the students, who could occasionally, as the commandant said, "play ball in the big league" of the U.S. military, but they could never actually win.

In almost every respect, the students' Washington agenda was fairly representative of those organized for other foreign military trainees who visited the capital from bases around the country. But there was one important exception: the SOA students visited Amnesty International, and appointments at human rights organizations did not figure into the sightseeing tours of foreign students from less controversial training establishments. Not surprisingly, the most contentious moment of the trip came when the group called at the offices of Amnesty International.

On the morning of the visit, the students gathered in the lobby of the Hotel Washington, where they were staying. Major Olsen, the coordinator of the CGS course, watched the men come down from their rooms and took note of stragglers. A marine with ramrod-stiff posture, Olsen wore wire-rimmed glasses and a heavily starched shirt that looked as if it could stand alone. He shaved what little hair remained on his head, and the result was a shining orb that glistened even in the subdued light of the hotel foyer. Olsen appeared to be in his thirties and was in excellent physical shape. He noted with a hint of pride that when the others returned the previous day from the Gettysburg battlefield, exhausted from the oppressive heat, he had headed to the hotel gym to work out.

The morning meetings were little more than standard overviews of the aims and activities of each organization that the group visited, but during a stop at the Interamerican Defense College—a think tank where some CGS graduates come later in their careers—one of the students asked the director about Amnesty International. The director, an army general, scoffed that "they are our best friends." He then threw up his hands and declared that there really was no alternative to talking to the human rights people. "Nobody controls the truth," he said, "and sometimes they attribute things to us that are not true." With this kind of an endorsement from a U.S. general, it was difficult to believe that anyone would treat the Amnesty visit seriously.

Lunchtime approached, and the students gathered in Hogart's Restaurant, where there was a sense that the afternoon session at Amnesty International would be unlike the morning's round of activities. As waitresses served what seemed like an unusual number of Diet Cokes (perhaps because of students' concern for their performance on a recent fitness test), the men at my table—two Peruvian naval officers, an Argentine, a Mexican, and a Colombian policeman—discussed their views of human rights organizations in anticipation of the upcoming meeting at Amnesty. The Mexican asserted that human rights organizations were really fronts for subversives. Others nodded their agreement. One of the Peruvians complained that only members of the state security forces were ever tried by international tribunals for human rights violations, and he noted the problems of Chilean General Augusto Pinochet. Guerrillas, he asserted, were just treated as common criminals. The conversation continued in this vein throughout the meal, and when we rose to leave, I asked the other Peruvian why he thought the SOA scheduled a visit to Amnesty International. He gave a sarcastic laugh.

Amnesty International's conference room was hot and stuffy. The air conditioning had stopped working on a day when the temperature outside reached ninety degrees, and nobody was happy. "They're violating my human rights," muttered one individual as he entered the room. Others speculated, seemingly in jest, that the absence of air conditioning was part of an Amnesty-initiated psychological operation against them. Over the next hour, the physical and emotional temperature in the room intensified further.

Amnesty representative Andrew Miller and two co-workers greeted the class and distributed informational materials to each student. Miller than gave a brief overview of the organization's activities and opened the floor for discussion. A barrage of hostile questions followed. Sánchez, a Bolivian with wire-rimmed glasses, demanded to know how Amnesty protected itself against infiltration by subversives. Claiming knowledge about the presence of subversives in Bolivian human rights organizations, he wanted to know if Amnesty investigated the credentials of people that it sent to Latin America. Other students asked how Amnesty supported its activities. When Miller replied that the organization did not take money from governments and only accepted donations from private individuals, a student asked for proof and echoed demands for information about the political backgrounds of donors. Rodríguez, a Mexican, asserted that drug traffickers, assassins, and terrorists supported human rights organizations and other nongovernmental institutions that always blamed human rights abuses on the army. A Peruvian demanded to know why Amnesty placed so much trust in civilians and their version of events. Why, he wanted to know, did it not work more with the armed forces?

Miller kept his cool. He described to students how he had witnessed uniformed soldiers murder civilians in the Colombian region of Urabá, and he insisted that numerous human rights organizations confirmed the massacre and the participation of Colombian soldiers in it. This anecdote incensed the two Colombians in the group. Lieutenant Colonel Héctor Peña claimed that accusations about the Colombian military's connections to paramilitary forces were completely false. Such alliances were isolated instances, he maintained, and they did not represent an institutional policy. These claims, however, were disingenuous. Peña was himself a member of an army battalion based in the highly conflictive Magdalena Medio region, where paramilitaries operated freely and coordinated operations with local military commanders (HRW 1996).

As the meeting finally wound down, Major Olsen asked the last question: what was Amnesty's position on the charges against the School of the Americas? Miller replied that the organization was currently involved in a study of military training around the world in an effort to construct a broad picture, and that it would assess the SOA within this context.[10] Throughout the charges and counterexplanations, my copious note taking drew nervous glances from one of the SOA staffers, and, after we left the building, he asked me what I thought about the event, insisting that it was an example of democracy in motion. Another staffer, who was less interested in my opinions, remarked that I had seen nothing compared to what happened the previous year at a different human rights organization where "there was almost a fist fight."

Jokes about human rights bounced from one student to another as the buses pulled away from the building, and even the laconic Major Olsen got into the act, echoing the earlier quips of others about the air conditioning. Doltish remarks about human rights peppered group banter for the remainder of the week, and on day four of the group's Washington visit, the question of human rights emerged again during a briefing at the Pentagon, where the SOA students met with Dr. Mary Grizzard of the Political Military Division's International Policy Integration and Assessment Directorate.

Grizzard began by asking people to introduce themselves and to express their questions or concerns. When it was his turn to speak, Major Félix Molina, a tall, mustachioed Bolivian, stood up and complained about Amnesty International. "Although it is a little vulgar [to say]," he began, "there are two young men [Miller and an associate] there earning a living by misrepresenting Latin America." He then asked Grizzard for Pentagon assistance in better controlling such people and organizations. Grizzard replied like the good public affairs expert that she was. Disappointing Molina, who had perhaps expected more support from an ally, she said that Amnesty was a nongovernmental organization over which the military had no control. Yet she advised the students on the importance of having a competent public relations person to deal with negative images of the armed forces. Such a person, she insisted, can always give the "good news."

Human Rights on the Homefront

To what extent, we might ask, were student views about human rights actually challenged by the SOA's program, despite its considerable shortcomings? And to what extent did any of the students begin to reflect seriously on the human rights records of their own militaries? All of the students did not conceptualize human rights in the same way, nor did everyone participate in the jokes and banter that followed the visit to Amnesty International. Some of the men engaged the human rights issues seriously in classroom discussions. Respect for the right to life and other basic freedoms, however, was not only an individual question; it was a social and political matter.

The SOA's forty-hour human rights program is just one arena in which ideas about human rights are formed. How rights are conceptualized, and whether they are respected or abused, depends to a considerable degree on the policies of particular militaries, the character of individual units within them, and the shifting balance of power in the countries from which students come, and to which they ultimately return. As mentioned above, most of the military institutions represented by CGS students were neither held accountable for human rights violations nor substantially reformed in the aftermath of war and civil strife, despite the efforts of human rights organizations to bring perpetrators to justice. It is from within these settings that the graduates of the SOA's human rights program develop their understandings.

The case of Honduran navy major René Maradiaga provides one example. Maradiaga finished the CGS class and went back to Honduras at the end of 2000 to become a human rights instructor. Six months after his return, he discussed his career and his understanding of human rights with me. What emerged was a view of human rights less intent on protecting Honduran civilians than on legitimizing the Honduran armed forces at a time when their mission was unclear and subject to debate.

Major Maradiaga began his military career in the 1980s, when Central America became a battleground for the Reagan administration's war on communism. Civil wars raged in Guatemala and El Salvador, and the U.S.-backed Contras, operating from Honduras, sought to destabilize the Nicaraguan Sandinistas. The United States spent millions of dollars on the Honduran military to convert it into a bulwark against leftist insurgency, and a human rights nightmare ensued in the region. Armies and military-backed death squads in El Salvador and Guatemala murdered peasants, students, labor activists, and church leaders, and thousands of people fled into exile. Tens of thousands of people died or disappeared. The violations attributed to the Honduran military were fewer by comparison, and a death squad within the army, known as Battalion 3-16, was responsible for most of the assassinations and disappearances (see chapter 3). During this period of turmoil, young Maradiaga led navy patrols around the Gulf of Fonseca to disrupt the flow of arms to the Salvadoran guerrillas of the Farabundo Martí National Liberation Front. Occasionally the patrols intercepted small boats carrying AK-47s, hand grenades, and other small arms, but the arms traffickers themselves, he maintained, always got away.

When Maradiaga reflected on these years, he drew on the Manichean cold war dichotomies that had assigned people to different and opposed categories, and that had, on more than one occasion, determined whether they lived or died. "I began my career," he explained, "at a time

> when communism arrived in Nicaragua. Our nation was being asphyxiated by the wars in [the region]. We share three frontiers with other countries, and they were all threatened by the spread of communism. We have a frontier with Nicaragua where the Sandinista Liberation Front had just taken power and proclaimed to the entire world that they were socialist communists. El Salvador was about to fall to the guerrillas. And we looked at Guatemala with its ancestral guerrilla movement. Our country was surrounded by bubbling caldrons. These countries, as the analysts used to say, were going to fall one by one, like dominoes.

Maradiaga distanced himself from the violence of the past and insisted that, as a navy officer, his hands were clean. He further claimed that the SOA's human rights course had an important impact on him. The course, he asserted, inspired him to address human rights in his final term paper, titled "Human Rights and Military Impunity," and his interest led to an appointment as a human rights instructor after his return to Honduras. Yet as Maradiaga expounded on his views about human rights, it became clear that his concerns focused less on the well-being and protection of Honduran civilians than on the integrity of the armed forces and the protection of their members. "Being sensitized to human rights," he said, "makes us appreciate the importance of the commands that we exercise, the importance of our authority and the authority that we delegate. [We bear responsibility] for the decisions that we make as commanders and whether these decisions bring adverse circumstances down on ourselves and the institution." He went on to describe his approach to human rights training in an advanced course for Honduran officers: "We are living in a new era, an era in which we do not need to open old wounds. What we want is to get rid of the wounds. And that is what we are inculcating in our soldiers—respect for human rights, a respect for the rights of others that does not undermine our rights [the military's rights]."

Maradiaga was, of course, well aware of efforts to bring military perpetrators to justice in Honduras and across Latin America. Truth-seeking commissions in a variety of countries, including Honduras, had published reports that implicated various militaries in human rights violations, and the threat, if not the reality, of prosecution troubled militaries deeply. Even though Maradiaga had never been charged or implicated in a human rights crime, he found challenges to military impunity an affront to the power that his institution had long enjoyed.

Maradiaga brought up the controversial visit that his SOA class made to Amnesty International nearly a year earlier. Despite the fact that, according to Maradiaga, Amnesty "works with information provided by anyone," the visit proved useful, he said, because it helped him understand how nongovernmental organizations operate. He explained that the encounter "made me appreciate how a small incident [in one of our countries] can have international repercussions as a human rights violation." He went on to describe how "this can happen when someone channels the information [to Amnesty]. It doesn't matter if it is analyzed or that it is reasonable, examined or true. A simple complaint can become an international human rights scandal for a country. [The complaint] could even be a dispute between two families. But Amnesty says, no, this is what is coming from your country, and we are wasting time trying to analyze it here. They will investigate it [in the United States]."

At the time of our conversation, the flow of U.S. military assistance to Honduras had slowed to a trickle. In the mid-1990s, following the end of the Central American civil wars and the fall of the Berlin Wall, Washington policymakers could no longer justify enormous expenditures on the Honduran military. Key domestic institutions, once controlled by the military, were under the management of civilians, who accused the armed forces of rampant corruption, and revelations about the atrocities committed by Battalion 3-16 had also tarnished the image of the armed forces. Faced with the loss of legitimacy, a drastically reduced budget, and the disappearance of a credible domestic "threat," the Honduran military was searching for a new mission and struggling to rebuild its public image. Maradiaga's remarks emerged from the problematic way that the Honduran military—like militaries elsewhere in Latin America—embraced a particular discourse about human rights to dissociate itself from the cold war past and to reassert the authority of the armed forces at a time when their purpose was unclear and subject to controversy.

The situation is more troubling in the Andean countries, where the United States has declared war on drugs. Enormous amounts of military aid for training and equipment are pouring into the region, where the violation of human rights has gone hand in hand with the strengthening of old, and the creation of new, security forces. Between 1997 and 2000, Colombia, Ecuador, Peru, and Bolivia accounted for over 60 percent of the students at the SOA. Although the SOA is by no means the only place where soldiers are trained by U.S. instructors, soldiers, as we saw in chapter 5, consider training in the United States more desirable and prestigious than pursuing the same course of instruction at home.

Earning the privilege to attend a U.S. military school, however, is tied to one's performance locally. In the Andean countries of Bolivia, Peru, and Colombia, security forces are under considerable pressure from the United States to demonstrate concrete results in the war on drugs and, in the Colombian case, the battle against guerrillas. "Progress" is measured in terms of the number of arrests, searches, and seizures, cocaine pits destroyed, and, in Colombia, by the dead bodies that the security forces produce. This production logic underwrites the career advancement of aspiring soldiers and antinarcotics policemen, and it influences the recruitment of trainees for admittance to prestigious foreign training schools, where students receive further instruction on the "management" and "application" of violence.[11]

The Colombian military provides the most egregious example of this process. It is involved in an expanding war with two guerrilla groups—the Revolutionary Armed Forces of Colombia (FARC) and the National Liberation Army (ELN)—that, since 1990, has left some thirty-five thousand Colombians dead. The Colombian security forces were initially responsible for most of the deaths, which included combatants and noncombatants, but as they have come under intense international pressure over the last five years to improve their human rights record, they have outsourced activities to paramilitary forces with whom they maintain alliances. The military claims that it has nothing to do with these groups, who now commit 70 percent of all reported massacres and extrajudicial executions, and it blames them for actions that it teaches and supports but does not condone publicly (HRW 1996). Not surprisingly, the paramilitaries have become more powerful and more deadly in recent years.

Yet in armed confrontations and in areas under guerrilla control, soldiers understand that maximizing the body count is a criterion for promotion and career advancement. The body count is particularly important for midlevel field officers—captains and majors—who command the units that engage rebel forces and who have a stake in career advancement within the armed forces. An army veteran I interviewed in a Colombian city lamented this practice because it had caused the army to lose prestige among civilians, particularly in rural areas. Majors, captains, and those under their commands, he claimed, killed people to be perceived as heroes and to receive promotions. They justified killing civilians by labeling them rebel collaborators, and they would place guns near the bodies of dead civilians to earn credit with superiors for killing more guerrillas. Sometimes, too, they ordered a military operation with the understanding that there would be no prisoners. He explained that a primary motivation for these practices was "to win a trip abroad" (*ganarse un viaje al exterior*).

"What do you mean," I asked, "a trip to someplace like the School of the Americas?"

"Exactly," he replied.

He went on to insist that he explained the problem to two generals for whom he had worked as an aide, but he was disappointed when they demonstrated little interest in his concerns.

Paramilitaries help soldiers reach their goals by supplying them with civilian corpses in exchange for weapons. Officers then dress the dead in combat fatigues and claim that they are guerrillas. The practice is referred to as "legalization" (Guillermoprieto 2000). High-ranking military leaders use the official figures of enemy dead, which appear daily in rightist Colombian newspapers, to convince weary urban readers of the army's progress against subversives and to justify the need for more military training. Human rights disappear under this brutal production logic.

All of this takes place in an environment of widespread impunity that is fortified by the United States. Perhaps the most egregious example of U.S. complicity is the approval by the United States Congress in 2000 of a $1.3 billion package of mostly military aid to Colombia. The legislation made a mockery of human rights; first, by strengthening murderous security forces and bolstering the position of hard-liners who support a military solution to the conflict, and then by adding the proviso that President Clinton could waive accompanying human rights restrictions for "national security" reasons.

In Bolivia, antinarcotics police forces operate on the basis of a similar production logic. The UMOPAR (Unidades Móviles de Patrullaje Rural) is charged with enforcing the eradication of coca in the Chapare region, and its operatives work less as a police force than an army unit. They are trained, equipped, and funded by the United States, and many of the policemen receive instruction at the School of the Americas. The force is responsible for the majority of human rights violations in Chapare, including arbitrary arrests, illegal searches and seizures, and the violent suppression of peaceful demonstrations. Yet the number of arrests and destroyed cocaine laboratories, as well as the maintenance of "order" in the Chapare, all count as evidence to U.S. drug war crusaders-in-suits that progress is being made. And, much as in the Colombian military, they also contribute to the career advancement of individual policemen and open opportunities for training at the School of the Americas. The UMOPAR is not the only militarized police force that operated in the Chapare, and it is not alone in its abusive treatment of coca growers. The so-called Ecology Police, which would have impressed George Orwell with its name, is another counternarcotics entity known for its aggressive tactics in the Chapare. Despite its name, the Ecology Police is less concerned with protecting the fragile subtropical ecosystem than with portraying the coca bush as a dangerous environmental toxin, and its operatives locate coca fields for destruction and protect army conscripts sent to eradicate them.[12]

The army placed CGS graduate Julio Miranda in command of the Ecology Police after it quickly promoted him to lieutenant colonel upon his return to Bolivia in late 2000. The posting placed Miranda in the middle of the United States' campaign to wipe out coca cultivation in the Chapare at a time when clashes between the security forces and peasant coca growers were turning increasingly violent. During the first year of his command, hardly a day passed in the Chapare without some form of peasant protest against the government's harsh coca-eradication policy (see chapter 7). Tensions ran particularly high in late October 2001, ten months after Miranda had returned. As they had done many times in the past, coca growers were threatening to blockade the major highway through the Chapare if the government did not modify its hard line against them, but the government refused to budge and sent thousands of soldiers to the region in a show of force. The Ecology Police was one of several militarized forces charged with imposing "order."

Truckloads of armed men moved up and down the highway in the days leading up to November 6, the deadline set by the peasant confederation for the government to meet its demands. Soldiers conducted early morning drills on the road, marching and chanting in exercises that were intended to intimidate local residents. They also patrolled the streets of local towns and set up camps along riverbanks and in jungle clearings. In some cases, communities were completely surrounded by military encampments. Men, women, and children boldly defied the security forces by organizing protest vigils in which groups of demonstrators marched out to the encampments, attempted to encircle them, and then peacefully observed the soldiers' activities. They carried the wiphala—a multicolored indigenous flag—with them. The Ecology Police, however, broke these vigils up with tear gas and confiscated the wiphalas.

The abusiveness prompted the Chapare's public advocate, Godofredo Renike, to visit Lieutenant Colonel Miranda in an effort to lessen tensions. Peasants charged that Ecology Police operatives were using the tear gas containers as lethal weapons by shooting them directly at demonstrators, and several peasants displayed ugly scars where they had been hit. Renike had already amassed a considerable collection of used tear gas canisters given to him as evidence by victims, and he collected still more from groups of protesters staging vigils along the highway, as he and I drove to the police headquarters.

A glossy brochure about the Ecology Police lay on a coffee table in Miranda's air-conditioned office. The booklet featured a picture of Miranda on the inside cover and a statement about the mission of the force that portrayed Miranda's men as the lonely, heroic defenders of law and order in the region. It asserted that "the members of the Ecology Police, separated from their loved ones, risk their lives to carry out their assigned mission with a firm commitment to the eradication of coca from the region."

Miranda greeted Renike and me cordially and ordered an aide to bring coffee. He showed me a picture of the CGS class that hung on a wall behind his desk, and we chatted about the current activities of some of the other Bolivians from the class. Renike then explained the purpose of his visit. He had not come to reclaim the impounded wiphalas but to encourage Miranda to ease up on the peasants. He pointed out that pressure on both sides was mounting as the negotiating deadline approached, and he suggested to Miranda that respecting the peasants' rights to fly the wiphala could calm tensions. Miranda listened politely to Renike's request and stated that both men should do everything possible to minimize potential conflicts. He insisted, however, that the peasants were flying the flags in inappropriate locations and provoking his men. Although Miranda never ceased to be courteous and gracious, it was clear that the issue of the wiphalas enraged him. The flags boldly proclaimed indigenous opposition to a racist state that had criminalized their most important crop, one that, peasants insisted, had deep roots in their culture. Addressing himself to me, Miranda commented about a fellow policeman whom peasants had allegedly injured so severely that he became a paraplegic. The man's wife and children had no means of financial support and were reduced to the care of sympathetic relatives. The unspoken message that emerged from Miranda's story of tragedy was that barbaric Indians, not the security forces, were violating the rights of the police, and any defense of them was illegitimate. And not surprisingly, the Ecology Police's excessive use of force and its practice of confiscating the wiphalas continued.

If Miranda had learned anything about human rights at the School of the Americas, it did not seem to apply to coca-growing Indians, who seemingly operated outside his universe of moral responsibility. Miranda's intransigent position on the flags probably did little to strengthen the Bolivian government's position in the Chapare or advance the U.S.-sponsored war on drugs, but it certainly provoked peasant antipathy to the state. Such uncompromising behavior typified the "zero tolerance" policies of the Bolivian government, which enjoyed the full support of the United States. Miranda was simply carrying out his duty, a role for which he had been groomed at the School of the Americas, and peasants' rights were clearly not his concern.

CHAPTER 7

✛

Disordering the Andes

COSECHA DE LA COCA
La coca se cosecha cuando está
bien madura. Después se hace secar.
Se plancha, se embolsa y se lleva al mercado de coca,
y la planta sirve para los caballeros para su trabajo,
y para mi vida y para mis estudios.
50 libras cuesta 500 bolivianos.
—María Isabel Vargas, 4th grade, Chipiriri, Bolivia

By the late 1990s, nearly half of the SOA's trainees came from the Andean countries of Peru, Bolivia, and Colombia. Following the fall of the Berlin Wall and the end of the cold war, the United States military had to look elsewhere for a credible threat to national security, as fighting communists in Central America soon became obsolete. The collapse of Soviet and Eastern European communism eliminated the most important justification for enormous military budgets, U.S. intervention abroad, and the School of the Americas, and the 2001 attacks on the United States had yet to fuel a new wave of militarization. Yet maintaining U.S. hegemony in Latin America required a new agenda for the armed forces. Pentagon strategists soon identified the burgeoning international narcotics traffic as a growing menace to U.S. security and wasted little time in declaring war on drugs in Latin America. The drug war provided a mission for the School of the Americas, and it was increasingly fought in the Andean region.

Although there was nothing new about the drug war,[1] the Andean initiative—a $2.2 billion economic and military aid program initiated in 1989—marked the shift from contending with communism in Central America to battling drug traffickers in South America (Tate 2001). Waging war on drugs provided a convenient rationale for the armed forces to argue against any reduction in their post–cold war budget and the scope of their involvement in Latin America, when cuts in defense appropriations, domestic base closures, and talk of a "peace dividend" were regularly in the news. Rising domestic fears of crime stoked the drug war, and demands to root out the problem "at the source" prompted lawmakers to point accusatory fingers at the Andean countries where most of the cocaine imported into the United States was manufactured. This chapter scrutinizes the local-level consequences of the drug war, neoliberal economic restructuring, and growing militarization for peasant coca growers in the Bolivian province of Chapare and the Colombian department of Putumayo, which are major cultivation centers for coca leaf, the basis of cocaine. It does so by exploring how impunity shapes the violent reconfiguration of economic and political relationships that is occurring in these regions.

The cocaine trade placed the Andean nations in a vexing dilemma. On one hand, it enabled them to occupy a competitive niche in an increasingly globalized economy, generated a hefty return on investment, and created employment locally, and it did so as free-market policies undermined local agriculture and industries and pushed more people into poverty. Cocaine profits eased economic pressures on central banks by bolstering hard-currency reserves, stabilizing exchange rates, financing imports, and reducing balance-of-payments deficits (Toranzo 1997), and the demand for workers to manufacture and transport a gooey cocaine paste, from which pure cocaine hydrochloride is derived, generated well-paid jobs locally. The expansion of coca also provided a livelihood for peasant families with few other economic alternatives.

On the other hand, the illicit drug industry represented a particularly vicious form of "gangster capitalism" that created enormous instability. The production of cocaine operated outside state controls that regulated capital accumulation and governed competition among capitalists. It required the development of extensive networks based on corruption to neutralize the state's efforts to suppress it, and drug dealers used violence as a primary means of conflict resolution (Uprimmy 1995). All of this threatened domestic tranquility, especially in Colombia. It discomfited foreign investors, who sought stable business environments, and it threatened the creation of the Free Trade Area of the Americas, a hemisphere-wide, free-trade project aimed at consolidating U.S. economic supremacy. Yet despite the mayhem major drug traffickers generated, the cocaine barons were not the primary casualties of the drug war.

The principal victims were peasant coca growers. Coca was the only crop that could be marketed in regions without stable roads, and it provided families with a modest living, as profits from its cultivation were sustained by the illegal drug traffic. By orienting agricultural production to the most lucrative export crops, peasants were doing exactly what neoliberalism prescribed. Yet states criminalized coca cultivation, and they required peasants to adopt a series of unviable alternatives or suffer the consequences. Coca cultivation exposed peasants to increasing levels of state violence that were unleashed by governments less concerned with addressing the social and economic difficulties of peasant families battered by free-market restructuring than with reducing the area planted in coca by almost any means necessary. Government officials turned the armed forces against growers to control the lawlessness and disorder that were to a considerable degree the outgrowth of their neoliberal policies. These policies, enacted in Bolivia in 1985 and then in Colombia in 1990, incorporated the countries more tightly into the U.S.-dominated global economic order by lifting tariff barriers on imported goods, relaxing restrictions on international capital flows, privatizing public companies, and cutting social spending. They also generated widespread unemployment, intensified social inequalities, and sparked an increase in common crime. As the turn to free-market capitalism created more instability that states could not control, it intensified the demand for better-trained militaries. Exploring the situation of peasant coca growers is important for appreciating the consequences of militarization and military training for ordinary people who find themselves increasingly targeted by the security forces.

The chapter first positions the Bolivian province of Chapare and the Colombian department of Putumayo within the broader context of global economic restructuring and the Andean drug war, as it unfolded in 2001 and 2002. It then considers how the dynamics of impunity shape the changing relationship between state security forces and deepening social inequalities in these regions. As we saw in chapter 6, impunity is constantly defended and regenerated at the SOA and among the armed forces of the Americas. It shields perpetrators and perpetuates the lack of accountability enjoyed by former military rulers, as well as armies, police forces, paramilitaries, and so forth. It is useful, however, to understand impunity as a broader phenomenon that extends from civilian institutions and the private sector to the repressive apparatus of the state.

Impunity reinforces a particular kind of "order"—political, economic, social—that is necessary for the expansion of free-market capitalism in Latin America. It allows dominant groups—government officials, financial institutions, multinational corporations—to enact destructive social and economic policies that fuel accumulation and, in the process, generate widespread misery. Impunity also enables them to enforce these policies by calling on the security forces to repress ordinary people who seek to hold them accountable, or whose daily struggles to survive threaten dominant notions of order. Impunity thus shields civilian and military officials from any negative consequences, and it leaves coercive entities intact and capable of generating further violence against defenseless communities.

For peasants and other impoverished peoples, the central feature of impunity is their inability to stem the violence directed against them or to hold perpetrators accountable. When experienced "from below," the political and economic violence of impunity generates disorder, because it ruptures social support networks, displaces entire communities, gives rise to vigilantism, and creates pervasive fear. All of this makes it difficult for people to reconstruct shattered lives, and it aggravates suspicion and hostility among peasants and working people, as well as between them and more powerful groups. People are not only more vulnerable to the terror and manipulation of dominant groups. They may also turn to vigilante justice, because the state either will not, or cannot, hold perpetrators accountable. In such contexts, victims may become victimizers.[2] Impunity is therefore closely tied to the growing social decomposition that accompanies the creation of a new global economic order.

My analysis first explores impunity for overt acts of state-sponsored violence, such as massacres, murders, and the destruction of property, perpetrated by state security forces and allied paramilitaries. This kind of violence is intended to impose a particular kind of political, economic, and social order desired by dominant groups, but it frequently gives rise to chaos and disorder that the state cannot manage. The discussion then considers how impunity places peasant families "outside the law" by denying them a political voice and the resources, such as economic security, health, and education, that are basic conditions of citizenship.[3] This aggravates poverty and forces people to grow coca simply to survive from one day to the next, and states face fewer restrictions in unleashing violence against newly criminalized populations, which reinforces the exclusion and marginalization of peasant families. Peasants must therefore contend with both the order and the disorder created by impunity-driven political and economic violence by constantly reconfiguring their social relationships simply to go on living. This is a volatile, highly unstable process that the state both creates and struggles to control, and it is why the maintenance of U.S. hegemony in Latin America requires huge military budgets and the maintenance of institutions like the School of the Americas.[4]

Chapare: Eradication and Alternative Development

If the recent history of the Chapare has been a tale of coca cultivation and drug trafficking, it is because the governments of Bolivia and the United States provided the basic infrastructure and entrepreneurial encouragement to support it. As early as the 1960s, state officials in Bolivia and U.S. development experts encouraged the settlement of Chapare's sparsely populated, subtropical lands between the highland valley of Cochabamba and the immense stretches of flat grasslands to the north, and they urged peasant settlers to produce cash crops for the market and become "modern" farmers. The United States provided the funding and the technicians to construct a road into the region, and it extended technical assistance for so-called peasant colonization projects. The vast majority of peasants who migrated to the region, however, did so on their own, through the formation of local-level organizations known as *sindicatos* that became synonymous with communities. The sindicatos organized social life in the frontier: they distributed land, planned communal work, mediated disputes, and initiated infrastructure projects in a region where the Bolivian state had little presence.[5] The settlers planted food crops—corn, rice, cassava—and coca, which became their primary cash commodity.

Coca is not cocaine. It is brewed as tea, chewed as a mild stimulant, and used for a variety of medicinal and ritual purposes. Peasants and indigenous people have grown it for centuries, and consumption of the leaves is legal in Bolivia. The coca bush is a hardy plant that grows well in the Chapare. It requires little attention, yields three harvests a year, and produces for up to twenty-five years if cared for properly. The dried leaves are easy to store, and their sale to peasant consumers in the highlands has long provided Chapare settlers with a modest cash income.[6]

By the late 1970s, however, soaring international demand for cocaine spurred a jump in the price of coca, and peasant growers found themselves riding the wave of an export boom fueled by drug users in North America and Europe. Narcotrafficking fit into the boom-and-bust pattern of the national economy, but as had been true of similar boom periods in Bolivia and elsewhere in Latin America, peasant families were not the primary beneficiaries of the rapid wealth.[7] An alliance between cocaine traffickers based in the lowland city of Santa Cruz and a segment of the armed forces facilitated the emergence of the illegal Bolivian narcotics trade and amassed most of its profits. Increasingly larger quantities of Chapare coca were siphoned into makeshift jungle laboratories, mixed with a cocktail of chemicals, and manufactured into a sticky paste that was purchased primarily by Colombian drug dealers (LAB-iepala 1982). The paste was refined into pure cocaine hydrochloride in Colombia and then smuggled into the United States. The area under coca cultivation in Chapare expanded rapidly, attracting new migrants in the wake of the collapse of the legal economy in the early 1980s and Bolivia's plunge into neoliberalism in 1985. Thousands of tin miners lost their jobs with the closure of state-operated tin mines, and, together with ruined peasants, they pinned their hopes of a better future on coca cultivation in Bolivia's subtropical valleys.

Cocaine replaced tin as Bolivia's most important—albeit illegal—export in the 1980s, when Ronald Reagan presided over a substantial expansion of U.S. antinarcotics activities, and George Bush's 1989 Andean Initiative increased the role of the United States in financing, training, and equipping police and military forces in the region. Under heavy pressure from the United States, the Bolivian congress approved Law 1008 in 1989, a controversial piece of legislation that created the juridical basis for waging war on the people of the Chapare. Law 1008 criminalized Chapare peasants with the stroke of a pen by abolishing their traditional right to grow coca, and it blurred the distinction between coca growers and drug traffickers, and between coca and cocaine (Farthing 1997). Peasant coca growers had suddenly become "outlaws" in the sense that they had literally been placed "outside the law," and if they continued to grow coca in the Chapare, where Law 1008 proscribed it, they would be treated as criminals by the state.[8]

Law 1008 not only criminalized thousands of coca growers, it restricted the individual rights of people accused of drug trafficking, who, in most cases, were low-level drug industry workers caught smuggling small quantities of paste. The law limited their right to defense and denied them provisional liberty. It also established an independent judiciary and a special, militarized, antidrug police force—the Fuerza Especial para la Lucha Contra el Narcotráfico (FELCN)—to coordinate the fight against drug traffickers. FELCN was an elite unit shaped by the U.S. Drug Enforcement Administration, and its members received higher salaries than ordinary policemen. The Unidades Móviles de Patrullaje Rural (UMOPAR) was a militarized rural entity that belonged to FELCN and operated in the Chapare, where it was responsible for the suppression of coca cultivation and the apprehension of cocaine traffickers. Many of its members received training at the School of the Americas, or from U.S. Army Special Forces teams that rotated through the Chapare, and it was widely feared by peasant families because of its abusive practices.

The flow of U.S. military assistance strengthened repressive security forces that operated in the Chapare when other state agencies—particularly those dedicated to social services—were withering under cutbacks prescribed by the International Monetary Fund (IMF). Not surprisingly, the military and new militarized police forces were the only viable entities through which state power was exercised in Chapare by the 1990s. Between 1997 and 2000, 287 Bolivian soldiers and elite counternarcotics police forces trained at the School of the Americas. Fifty-two percent of them took five courses: Anti-Drug Operations, Cadet Leadership, Medical Assistance, Military Intelligence for Officers, and Instructor Training. While the SOA emphasized officer training, the Special Forces sent small mobile teams of U.S. trainers to Bolivia to instruct low-ranking soldiers and policemen in basic combat skills and patrolling techniques, and to conduct joint training exercises with the Bolivian military.[9] The two training entities coordinated with each other, and U.S. instructors moved easily between them.

Colonel Roy Trumble, a former SOA commandant (1995–99), was part of a Special Forces unit in the 1980s that was based in Panama and began the training of South American counternarcotics forces. His unit trained the Bolivian UMOPAR and a widely feared Peruvian special forces unit known as the Sinchis. Trumble explained:

> We [the United States] decided that we were going to train a police element that had military capabilities . . . so we went and built a camp in the middle of the coca producing region of Bolivia. So we built this camp and took basically policemen and taught them to be soldiers, because the nature of the insurgent or the value of drug interdiction, required the same military techniques and tactics. Instead of doing a traffic stop, we [taught] them to find labs . . . how to use intelligence that we provided them, [and] helped them to plan operations. So we trained them to have an operational capability. We built the camp in the middle of the coca-growing region in the hope that the mere presence of this camp would have an impact on the drug traffickers.

The impact of the base on drug dealers who typically live outside the Chapare was considerably less than on the thirty thousand resident peasant families whom Colonel Trumble misrepresented as drug traffickers. These families resented the militarization of the drug war, the presence of the base, and the intensifying campaign to destroy their only viable source of livelihood.

Today, Special Forces instructors rotate on three-month schedules through the sprawling military compound outside the Chapare town of Chimoré. The rusty hulks of battered tracks, old cars, and minibuses litter a clearing near the main gate, where they erode in the tropical heat and humidity. Weeds and tall grasses entwine their fenders. Seized from alleged drug traffickers by UMOPAR, these decrepit specimens await the day when they will be sold at auction. The absence of sport utility vehicles and other luxury models attests to the fact that major drug traffickers rarely venture into the Chapare, and important drug lords—or even midlevel operatives—are not represented among the approximately two hundred and fifty prisoners held in a jail for civilians on the base. The prisoners are overwhelmingly peasants and poor people apprehended for minor crimes under Law 1008, or they are innocent bystanders caught in the wrong place at the wrong time. Despite the presence of a prosecutor's office next to the jail, the detainees are frequently held without charge for long periods, and even if they are eventually released, the law provides no compensation for the time spent in prison.

Not far from the jail and the prosecutor's office in a barrack-style building is one of the base's most bizarre features. A makeshift drug-war museum stands as a surreal testament to the labors of the armed forces in the Chapare. On the day that two companions and I visited, in June 2001, an UMOPAR policeman named Layme escorted us around. Layme showed us a series of disorganized exhibits that described the process of manufacturing cocaine and that displayed the people and arms captured by UMOPAR. In one exhibit, dirty plastic bottles of precursor chemicals surrounded a pit with coca leaves, and in another, photographs of disheveled prisoners were stuck to a board. Other exhibits featured old guns and pistols taken from alleged traffickers, a relief map of the Chapare, and various devices used to smuggle cocaine, such as old shoes, hollow bedposts, and truck bodies.

Neat white exhibit labels printed in English and Spanish were the strangest aspect of the museum. "Cocaine pit," read one. "Coca leaves," stated another. Who, we asked Layme, was the intended audience for this drug-war showroom, located in the midst of thousands of peasants who only entered the base as prisoners, had no knowledge of English, and often spoke Spanish as a second language? If any of this seemed odd to Layme, he gave no indication. He explained that the museum's purpose was to educate delegations of visiting dignitaries, and he noted that a group of international drug czars, who had recently convened in Cochabamba, had come to the museum on a postconference tour a few days earlier.

Layme was a young man from the highland city of Oruro and he lived on the base with his family. After completing a year of obligatory military service, Layme worked as an Oruro city policeman until UMOPAR recruited him in 1997 with an offer of wages substantially above those of an ordinary street cop. He also received additional training and instruction from the U.S. Special Forces. Layme seemed to get a perverse pleasure from his work, and he denied any qualms about destroying the fields of people who came from backgrounds similar to his own. The work, he said, was fun (*divertido*). He then posed in a slight crouch, raised his hands slightly, and talked about hunting down drug traffickers and their cocaine pits in the jungle. He went on to say that members of the Special Forces, whom he and his colleagues referred to as "the friends" (*los amigos*) because of their hard-to-pronounce English names, accompanied UMOPAR troops on nighttime raids from time to time. Unaware that this activity was illegal, Layme explained that the men did so because they got antsy in the barracks.[10]

Figure 7. Peasant leader Evo Morales confronts militarized policemen in the Chapare. Photo by Jeremy Bigwood.

By the time that we had this conversation with Layme, the drug war in the Chapare had been intensifying for a number of years. Many peasants felt that pressure to eradicate coca was just the first step in a concerted effort by the Bolivian government and the United States to dislodge them from their lands and remove them from the Chapare. They believed that the smog-choked United States coveted the region for development as an ecotourist mecca, and they did not have to look far to understand that the Chapare was being transformed in ways that completely disregarded their presence. As one entered Villa Tunari—the first major town in the Chapare—an enormous pink arch stretched across the highway and emblazoned on it was the greeting "Villa Tunari—Welcome to the Ethno-Eco Paradise." The paradise, however, was an imaginary, fool's heaven, where the "ethno" referred to indigenous peoples long driven from the region, and the "eco" suggested a tropical Shangri-La devoid of combative peasants. A partially completed luxury hotel loomed out of the dense vegetation farther along the road. Other hotels catered to the needs of well-heeled foreign and domestic tourists who came to the region on short visits when the political situation permitted, and these visitors occupied rooms that local people could never afford. Both the sign and the hotels suggested a vision of the Chapare's future that did not include the coca growers, and when I commented about the garish public greeting to a former Chapare resident, he quipped rhetorically, "What must the tourists think of paradise when Evo erects a road blockade?" He was referring to Evo Morales, the charismatic leader of the coca growers, and the road blockades that were a frequent form of peasant protest against government anticoca policies.[11]

Peasants' deep-seated fears about land loss were also inflamed by controversial 1996 agrarian reform legislation known as the INRA law, which sought to define individual property rights. Coca growers viewed the law as an unwanted intrusion by the state. It seemed to violate the notion, upheld by the Bolivian constitution, that "land is for those who work it" and to replace it with the maxim "land is for those who can pay for it" (Potter et al. 2002). Many peasants saw evidence in the law of the state's intention to undermine the authority of the sindicatos, which validated land claims at the time of settlement and organized the current resistance to coca eradication. The INRA law also required impoverished households to spend money on surveys and titling procedures, and peasants feared that if they handed over their land documents to the government, evidence of their claims would be lost forever.

It should come as no surprise that, over the years, the coca growers staged constant protests against government pressure to eliminate coca. Periods of negotiation alternated with outbreaks of conflict. The government initially exhorted peasants to "voluntarily" eradicate their coca and then to incorporate themselves into a U.S.-sponsored alternative development program. Alternative development, however, did not impress many people, even though for a brief period it compensated some families for the voluntary destruction of their coca fields. Peasants were eager to adopt reasonable substitutes for coca to avoid escalating government repression, but alternative development programs were less about alleviating poverty than coercing peasants to forsake their only viable cash crop. Like cold war civic action initiatives of the past, the entire alternative development program—whatever its good intentions—was subordinated to a coercive drug war strategy that sought the elimination of coca by intimidation. The conditions that would make alternative development successful—stable roads, credit, technical assistance, and high prices for substitute products—did not exist, and peasants who wished to replace their coca with other crops—pineapple, passion fruit, palm heart, and black pepper—had to dissociate themselves from the sindicatos and join parallel local-level organizations called associations (*asociaciones*) to receive credit and technical assistance.

The associations are a mechanism to isolate coca growers and create divisions among them through the selective distribution of financial and technical assistance. They exist because the United States embassy and the U.S. Agency of International Development (USAID) refuse to disperse alternative development funds through the seven municipalities of the Chapare, as mandated by Bolivian law. The democratically elected mayors of these municipalities all belong to the Movimiento Hacia Socialismo (MAS), the political arm of the Six Federations of the Tropics of Cochabamba. The Six Federations is an umbrella organization that consists of six regional federations of local sindicatos. It represents some forty thousand peasant coca growers, and it is the most powerful entity through which peasant opposition to alternative development programs and coca eradication is articulated. Although European development agencies work with the coca growers and their local representatives, technicians contracted by USAID operate through two hundred forty associations that represent some nine thousand families (Potter et al. 2002). Yet because of their inability to effectively establish alternatives to coca cultivation, association members are often dissatisfied with the experiences in these organizations and drift back to the sindicatos. Peasant leaders of the Six Federations generally take a pragmatic view of this practice and do not sanction those who seek assistance through the alternative development program. The result is that peasants tend to move between the sindicatos and the associations.

The ongoing confrontations between peasants and state security forces turned more violent after 1997, when Hugo Banzer assumed the presidency for the second time. Banzer and U.S. drug war zealots determined that voluntary eradication had been a failure, because peasants had simply taken the money offered in compensation for uprooted coca fields and replanted the crop elsewhere. Banzer declared a new policy of zero tolerance for coca cultivation in the Chapare. Peasants who refused to destroy their coca bushes faced the prospect of soldiers entering their fields and eradicating the coca, and this began to happen in 1998. Under the protection of heavily armed UMOPAR operatives, hundreds of army conscripts fanned out onto peasant lands from access roads that penetrated the region. Others flew into the most inaccessible places by helicopter and set up camps in soccer fields or jungle clearings, where they remained for

Figure 8. Peasant coca growers of the Six Federations debate whether to blockade roads throughout Bolivia. Photo by Jeremy Bigwood.

days uprooting coca bushes. According to numerous human rights observers, the soldiers terrified the local population and committed a number of abuses, including the excessive use of force, illegal searches and seizures, torture, and arbitrary arrests.

Peasants from the sindicato Guadalupe denounced the violations committed by security forces to Radio Soberana, the voice of the Six Federations, but few, if any, policemen and soldiers were held accountable because of the pervasive impunity enjoyed by the military. Elizabeth Jaldín stated that the soldiers "asked us where our husbands were. They threatened us, but my husband was in Cochabamba. They said that I was a liar and pointed a gun at me. You're going to talk, they said. We were afraid [to remain in the sindicato] and went into the jungle with our children. They hit the children to make them talk." Another individual described how he was working in his field when the soldiers arrived. "We were weeding the cassava," he said, "and in that moment, the eradicators arrived and shot three bullets into the air. They tied our hands behind our backs with our own belts. We were like thieves. Then they took us back to the house to pressure us to tell them who had arms." The soldiers robbed another woman and destroyed what they did not take. "They threw out

Figure 9. Peasants relaxing after a night of preparing road blockades.
Photo by Jeremy Bigwood.

Figure 10. Militarized policemen confront coca growers on a Chapare bridge. Photo by Jeremy Bigwood.

three quintales [hundred weights] of rice," she explained, "and they took buckets, mattresses, and clothing. When my daughter told them to stop, they told her to shut up. My house looks like a pig pen—everything is smashed and broken."[12]

U.S. officials declared the eradication operations successful at the end of 2001 and gloated that coca acreage in the Chapare had dropped 48 percent, from 38,000 hectares in 1998 to 19,900 hectares in 2001.[13] Two hundred fifty million dollars had also been removed from the economy of South America's poorest country, and many peasant families found themselves in a desperate situation. Yet despite all the claims of success, the campaign to impose a coca-free social order was a pyrrhic victory. The Bolivian government and U.S. development technicians neither incorporated peasants into new productive arrangements as the failure of alternative development made clear, nor insured against a resurgence of coca cultivation in the absence of a large military presence. They were left to confront the social and economic turmoil that the war on drugs had created, and as peasants regrouped and devised new strategies of struggle, families continued to defend their right to grow coca.

Perhaps more disturbing to the standard bearers of the drug war, however, was growing evidence that even though aggressive repression had reduced acreage in Bolivia—at least temporarily—and produced similar results in the Upper Huallaga Valley of Peru, production was shifting to Colombia, which soon replaced Bolivia and Peru as the primary producer of coca leaves. The southern Colombian department of Putumayo soon became the new epicenter of the drug war crusade.

Putumayo: Fumigation and Counterinsurgency

The much-heralded success in reducing coca in Bolivia and Peru took place as a forty-year-old civil war escalated in Colombia, and it shifted the balance of power in the U.S.-sponsored drug war. While Colombia accounted for only 12 percent of the coca acreage in 1994, the figure jumped to 71 percent, or 635,000 acres, in 2000 (Potter et al. 2000). Nine of the thirteen municipalities in Putumayo department contained coca, and by early 2002, nearly 300,000 acres of bright green coca bushes spread across the rolling hills and fields of the department. Several groups of armed actors—the U.S.-financed Colombian military, paramilitaries allied with the army, and guerrillas—contended for control of the region in what was at times a drug war and at times a counterinsurgency war that the United States refused to name.

The physical geography and the recent history of Putumayo share many common characteristics with the Chapare. The department occupies a series of undulating valleys and foothills that extend from the base of the Andes to the tropical plains of the Amazon basin, and it is an area where the presence of the central state has been minimal or nonexistent. Unlike the Chapare, however, Putumayo has a long contemporary history of political and economic violence that extends from an early-twentieth-century rubber boom in which thousands of indigenous people died (Taussig 1987) through the development of oil in the 1960s to the present coca economy. Many of the department's current residents are peasants from the highland departments of Nariño, Pasto, and Huila who migrated to the region in the 1960s and 1970s in search of land and, in some cases, fleeing persecution. The current violence that envelops Putumayo is not always a new experience for them. Other residents have arrived in the last several years, following the 1990 opening of the Colombian economy to greater global investment and the increased flow of foreign goods.

Free-market policies negatively affected domestic producers who could not compete with multinationals. Small-scale agricultural producers were hit particularly hard, as imported basic grains, such as corn, undercut the price of their own crops in local markets. The subsequent economic ruin of small-scale agriculturalists was less the result of fair competition than the oligopolistic market power of multinational corporations and high U.S. and European subsidies to domestic farmers that permitted crop dumping in places like Colombia. Unemployment soared to 30 percent by 2000, and, as in Bolivia, the coca boom offered a last-ditch hope to those ruined by unrestrained capitalism. It tempted people from both urban and rural areas to test fate and risk their lives in Putumayo, where over half of the 350,000 inhabitants of the department depended on coca cultivation by 2000.[14] Yet the coca boom also prompted many Colombians to turn to various armed groups—guerrillas, paramilitaries, and criminal gangs—for protection and survival, which had not been the case in Bolivia.

The dramatic upsurge in Colombian coca production and the growing power of guerrilla insurgents, who financed operations in areas under their control with a tax on peasant coca growers, deeply disturbed policy-makers in Washington. Earlier victories in the Colombian drug war, such as the dismantling of the Medellín- and Cali-based drug cartels in the 1980s, had amounted to little. The flow of cocaine into the United States continued. Smaller, more numerous drug-trafficking organizations re-placed the cartels and developed new smuggling routes through Central America and the Caribbean. To contend with the deteriorating situation, the United States made Colombia the largest recipient of military aid in the Americas with the passage of Plan Colombia in 2000. The $1.3 billion mostly military aid package included support for the training of three new counternarcotics battalions by instructors from the Special Forces and the School of the Americas. These battalions brought the drug war to Putu-mayo as part of a U.S.-orchestrated "push into southern Colombia." Lieu-tenant Colonel George Ruff, an intelligence specialist from the School of the Americas, helped train one of the battalions in Colombia. Ruff ex-plained that "by doing it down there, you truly have a captive audience focused on their problem. If you have it up here [at the SOA], you have students from a number of different countries. There is the benefit of, say, the Bolivians pointing out how they do it. Down there you say, this is the problem in Columbia, so you can focus your teaching."

Yet Putumayo—like the Chapare—was not the home to Colombia's wealthiest drug dealers, nor were Putumayo and neighboring southern departments the center of the cocaine traffic. The major smuggling routes and the paramilitaries who controlled them lay in the north, but U.S. strategists and their Colombian counterparts were less interested in cocaine traffickers and paramilitaries than in limiting the growing financial, military, and strategic power of leftist insurgents. Since the cold war, they had viewed the guerrillas as threats to U.S. interests and the domestic status quo in Colombia, and concerns about them prompted the initial militarization of the drug war by the first Bush administration. U.S. officials understood that the Colombian government would only commit its military to the drug war if it could use U.S. aid to fight the guerrillas (National Security Archive 2002),[15] and Plan Colombia supported a similar political project.[16]

Although several left-wing insurgent groups operated in Colombia, the Fuerzas Armadas Revolucionarias de Colombia (FARC) was the oldest and most powerful revolutionary movement, and its increasing strength at the end of the twentieth century provoked the greatest concern among U.S. and Colombian officials. The FARC had long operated in southern Colombia. It emerged in the mid-1960s from peasant self-defense groups that were organized by displaced highland small holders who sought to protect claims to land in newly opened frontier zones.[17] The guerrillas, however, did not become a major force until the 1980s, when they formed the Unión Patriótica (UP) and entered the arena of formal party politics, which allowed them to gain a foothold among the urban working and middle classes. The FARC's urban ties were subsequently severed by a paramilitary "cleansing" campaign that virtually eliminated the UP. Over a seven-year period, paramilitaries murdered over 2,000 UP members, including two presidential candidates. The wave of assassinations left the FARC isolated with its peasant base in the countryside. Its rural isolation was aggravated when the guerrillas broke with the urban-based communist party in the late 1980s, and many leftist intellectuals rejected the FARC's ideology and tactics after the collapse of the Soviet Union (Vargas 1999; Richani 2002). As coca cultivation expanded and Colombian cocaine traffickers consolidated their position in the world market, the FARC began to protect peasant growers from abusive drug lords and raised money for their own operations by taxing coca cultivators, cocaine laboratories, and drug dealers who operated in areas under their control. Together with revenues raised through extortion and kidnapping, the guerrillas doubled their forces to 20,000 troops and increased their military power, even as their political power waned and their long-term objectives became less clear (Vargas 1999).

Yet the FARC's growth was not only due to its ability to generate resources locally and build military muscle. The brutality with which government security forces and allied paramilitaries murdered, tortured, and displaced noncombatant civilians drove many people into the guerrilla ranks, and the FARC took advantage of the government's refusal to stem a deepening agrarian crisis (Vargas 1999) that was aggravated by a harsh "counter agrarian reform." Newly rich drug lords acquired between 3 and 4.4 million hectares of fertile, resource-rich land in key areas around the country, such as the Middle Magdalena region. In their struggle to consolidate property rights, they were joined by multinational corporations and the old agrarian elite—cattle ranchers, coffee growers, and agribusiness groups. There was nothing new about land conflicts in Colombia. Yet decades-long struggles were fueled with new intensity by cocaine profits and the open-door economy in which land was valued less for its productive potential than for the speculative opportunities embodied in it. New road and canal projects, for example, triggered speculation in areas that surrounded them, and land situated near oil fields and mineral reserves became the sites of conflicts to secure claims to it. As old and new sectors of the dominant class renegotiated their positions within a more globalized economy, they drew on an old Colombian pattern of contracting paramilitaries and building private armies to dispossess peasants, squatters, and guerrillas through the use of selective assassinations, massacres, and systematic terror (Richani 2002). By the end of the 1990s, nearly two million Colombians had been displaced from their homes.

U.S. propagandists coined the term "narcoguerrilla" to delegitimate the rebels, as the old ideological chestnut "communist subversive" became outdated, but narcoguerrilla was a typically misleading label. Even though the FARC benefitted from coca production and processing, it did not control the major national and international cocaine processing networks. The narcoguerrilla concept blurred the identities and goals that distinguished the FARC from other guerrillas groups, such as the Ejército de Liberación Nacional (ELN), and from the drug traffickers. It also obscured the involvement of the paramilitaries, sectors of the military, and members of the political establishment in the drug traffic.

The FARC and peasant coca growers were the main targets of the U.S.-

sponsored "push into southern Colombia" that was financed by Plan Colombia and facilitated by paramilitary attacks in Putumayo. The paramilitaries were groups of armed men, who included former and active-duty military officers, common criminals, and unemployed youth attracted by the relatively high wages. They ceased to be a local phenomenon in 1997, when disparate organizations federated under the umbrella of the Auto Defensas Unidas de Colombia (AUC) and coordinated a national strategy to eliminate the growing threat posed by the guerrillas. Its leader, Carlos Castaño, was formerly an enforcer for the Medellín drug cartel, and in certain parts of the country, such as Putumayo, the AUC maintained a close relationship with the institutional armed forces.

Known as a "sixth division" because of their ties to the Colombian military, the AUC wanted to claim the lucrative profits that flowed from the coca-growing region of Putumayo and timed its move into the department with the military. Hit squads from other regions entered towns and villages to assassinate suspected guerrilla collaborators and coca paste merchants. The army's Putumayo-based Twenty-Fourth Brigade collaborated with the AUC to assault their common enemy, the guerrillas, and the AUC, which had amassed considerable wealth through drug trafficking and money laundering, paid army officers for their collaboration.[18] At the same time, members of the U.S.-supported counternarcotics battalions worked closely with the Twenty-Fourth Brigade, using its facilities, intelligence, and logistical support and associating with individuals who supported the paramilitaries (HRW 2001). Meanwhile, Washington policymakers said relatively little about the AUC. Although the State Department eventually listed the paramilitaries, along with the FARC and the ELN, as terrorist organizations and agreed with human rights groups in blaming paramilitaries for most of Colombia's human rights violations, State Department officials largely ignored the AUC's links to the Colombian military. The AUC was therefore relatively free to expand its power and to murder ordinary Colombians with impunity, as the war targeted the FARC and peasant coca growers.

Throughout 1999 and 2000, the paramilitaries gradually wrested control of Putumayo towns and villages from the FARC, leaving a trail of death and destruction in their wake.[19] Their power, and the impunity with which they operated, was evident when I visited the little town of La Dorada with a delegation of human rights activists in early 2002. We passed through two military checkpoints—one operated by the Colom-

Figure 11. SOA graduate Colonel Mario Montoya commanded the Twenty-Fourth Brigade in Putumayo. Courtesy of Linda Panetta—Optical Realities / SOA Watch Northeast.

bian army, the other set up by an elite U.S.-trained army counternarcotics battalion—before entering the town square less than two miles past the last roadblock. Our driver stopped in front of the church, and a young man dressed in civilian clothing emerged from an adjacent building to greet us. He approached our vehicle before we could get out and identified himself as a member of the AUC. He informed us that the AUC controlled La Dorada and assured us that we had nothing to fear. The AUC was there to serve us. Such brazen acts could not possibly escape the notice of the army, nor take place without its complicity. A local resident who was forced to flee later recounted how the AUC had taken over the town. "There were large numbers of disappeared people" (*Había gente desaparecida en cantidad*), he recalled, "and the psychological stress was unbearable. People were afraid to leave their homes, and one day a shootout erupted in front of the church while the priest was saying mass. Nobody knew who was shooting at whom but it terrified people and heightened the state of anxiety in which they lived."

Paramilitaries also controlled the town of El Placer, where they operated a base and were identifiable by the large walkie-talkies that hung from their hips, and they seemed to run Mocoa, the departmental capital, as well. Although busy restaurants and pedestrian-filled streets provided a veneer of normalcy around the Mocoa town square, the appearance of tranquility was deceptive. "Yes, it seems very normal," a human rights observer said to me, "and then suddenly it isn't." I appreciated his point when my companions and I left a Catholic church, after meeting with the parish priest, and three men approached us on the church steps. They represented themselves as displaced peasants who lived in a refugee settlement somewhere on a hill that one indicated with a vague wave of his hand. Their clothing, however, was much too expensive for displaced people who had lost everything, and they were too interested in what we had done, where we were going, and whom we planned to meet.

The paramilitary takeover of Mocoa, El Placer, La Dorado, and other small towns in Putumayo coincided with the U.S.-sponsored aerial fumigation of coca fields in the countryside, where the FARC remained in control. Fumigation was tied to a coercive alternative development scheme in which peasants were forced to sign "social pacts" with the Colombian government that were designed to channel U.S. aid to those who agreed to manually eradicate coca. Some 37,000 growers consented in 2000 to destroy their coca within six months after the arrival of financial assistance or face the fumigation of their fields with herbicide, but they soon discovered that financial aid was either slow to arrive or never came. To make matters worse, adhering to the terms of the government's pacts was no guarantee against fumigation.[20]

The first round of aerial fumigation took place between December 24, 2000, and January 2, 2001, and according to the U.S. embassy, it destroyed 30,000 hectares in the municipalities of Orito, San Miguel, and Valle de Guamuez. The Catholic Church, however, stated that coca represented only 15,000 hectares; the rest of the area was planted in food crops. In November 2001, spraying began again in San Miguel and Valle de Guamuez, even though little or no government aid had reached pact signers who remained in compliance with the eradication agreements.

The experience of one peasant is indicative of the level of destruction suffered by rural families. Julio Meza[21] signed a pact with seventy members of his community, but because he never received the promised development aid, Meza borrowed $12,000 and planted a hectare of black pepper to offset the imminent loss of his coca. Then he stuck a white flag in the field to signal that the black pepper crop was legal. When the spray

Figure 12. Street mural shows the effects of fumigation on humans, food crops, and wildlife in Putumayo. Photo by author.

planes returned on November 24, 2001, they destroyed all the pepper plants, as well as fruit trees, coffee bushes, cassava, corn, and pasture. Two months later when I visited the family, their fields were covered with dead vegetation. Barren trees, withered corn husks, shriveled cassava plants, and desiccated pasture colored their farm and the fields of their neighbors in shades of brown and beige. Lifeless fish ponds stagnated in the tropical heat. Cattle that once grazed in verdant fields had either been sold or moved elsewhere. All of this raised disturbing questions about the real aims of Washington's fumigation program in guerrilla strongholds. As one U.S. embassy official who wished to remain anonymous told me, the reduction in the total area under cultivation—not crop substitution—was the cornerstone of U.S. policy. He added, "It may be best that people simply abandon the area or make their own decisions about how to survive. . . . People need to understand that the [U.S.] government absolutely will not tolerate coca cultivation."

When confronted with evidence of food crop destruction, embassy officials explained that, although aerial spraying was supposed to be done from thirty feet up, the presence of guerrillas forced planes to fly higher, which caused the herbicide to drift onto neighboring fields. But although drift might explain some of the damage, it did not account for the level of destruction in Putumayo. One Colombian expert on alternative development who wished to remain anonymous noted that "the only explanation is that fumigation was intentional," calculated to destroy peasant morale in an area under FARC control and to displace rural people to other areas of the country. In fact, displacement had long been a strategy, not an effect, of Colombia's civil war, and the fumigation campaign unfolded in a context in which the Colombian government could not defeat the guerrillas militarily. All of this was fueled by the political and economic violence that defines impunity.

Getting Away with Murder

Militaries, paramilitaries, and counternarcotics police forces enjoy astonishing impunity in Colombia and Bolivia, and in the absence of any accountability for brutal acts of violence, very little impedes their continued violation of the rights of others.[22] This, in turn, generates deep social fragmentation, insecurity, and exclusion, which give rise to forms of violence, such as vigilantism and common crime, that are not directly political in nature. States typically deal with the growing disorder and rising levels of violence by resorting to repression, which further disorganizes daily life at the bottom of the social hierarchy and shores up a dominant order that is intolerable for the majority of poor Bolivians and Colombians.

There is no accountability in Colombia, where forty thousand Colombians died as the result of political violence in the 1990s. Paramilitaries, supported by the armed forces, were responsible for over 70 percent of the deaths, and few suffered any consequences for their actions.[23] The violence wracked Putumayo with particular ferocity; in 1999, for example, paramilitaries committed thirteen massacres that killed seventy-three people. Because of their growing strength, and the mounting pressure on the institutional armed forces to clean up their human rights record, the paramilitaries increasingly waged a dirty war on behalf of the Colombian army. The military-paramilitary relationship was based on what Human Rights Watch called a "strategy of impunity" in which "supposedly 'phantom' paramilitaries that the military claims it can neither identify, locate, nor control take the blame for massacres and forced disappearances, allowing the military to evade responsibility . . . paramilitaries take the brunt of criticism for tactics taught, employed, and supported by the armed forces, but which they do not openly endorse" (HRW 1996, 61).

In the rare event that the case of a member of the paramilitaries or the armed forces reaches a Colombian court, judges are threatened and even murdered if they investigate it. Fearing for their lives, many magistrates simply hand the matter over to military tribunals, where it is typically dismissed. There are ninety-seven Colombian graduates of the School of the Americas who have been linked to human rights violations by multiple international human rights organizations,[24] but because of the high levels of intimidation experienced by members of the judiciary, the vast majority of these alumni have received promotions instead of punishment.

Bolivia is relatively free of the homicidal violence that has devastated Colombia, but impunity in the armed forces is widespread. The Bolivian Permanent Human Rights Assembly has documented fifty-seven peasant deaths by security forces since 1987, and none of the perpetrators has been held accountable (Ledebur 2002, 15). One of the dead is peasant leader Casimiro Huanca, whom I interviewed a few months before his death.

Huanca led the Chimoré Federation of Coca Growers. He was an Aymara peasant, born in the Bolivian highlands, who had come to the Chapare in search of a better livelihood. The U.S. embassy identified him as a "die-hard cocalero in Evo Morales' camp."[25] As we sat in the head-quarters of the Chimoré Federation in June 2001, surrounded by a group of cocaleros, Huanca denounced the crop substitution program in the Chapare, asserting that the money for alternative development simply disappeared into the pockets of technicians contracted by the U.S. Agency for International Development. More importantly, he indicated that there were no markets for alternative crops. Bad roads impeded the movement of products from the fields to urban markets, and the government's free-market policies undercut the ability of Chapare peasants to sell their produce. Six months later, during a peaceful demonstration against alter-native development, Huanca was fatally shot by a member of the Chapare Expeditionary Force, a counternarcotics entity created by the United States that was sent to break up the protest.

The military immediately placed the blame for the shooting on "drunken, rock throwing" peasants, but after reviewing a videotape of the events filmed by a local cameraman, U.S. Defense Department officials stationed in Bolivia determined that the military's version was "inaccu-rate at best and a complete fabrication at worst." Nevertheless, a Bolivian military tribunal found that the soldier who shot Huanca did so in self-defense, and it did not punish him. It then concluded that his commander, an army lieutenant, had failed to exercise sufficient control over the troops and confined him to his quarters for forty-eight hours. The tri-bunal's behavior was so outrageous that even U.S. officials stated that the findings "rubberstamp and thereby vindicate the military's own version of what occurred."[26]

The failure to hold members of the security apparatus accountable emboldens repressive actors who do not have to fear any negative conse-quences for themselves. Military and paramilitary violence is often exem-plary and intended to paralyze people with fear. This is what happened in Putumayo, in the little town of El Tigre, on January 9, 1999, when armed men massacred nearly thirty people, slit open their bodies, and threw the corpses off a bridge into the river. A year later, El Tigre remained in the grip of terror. As I drove over the same bridge, a dead body lay on the river bank, surrounded by a group of people. The selective murder of individ-uals and the public display of their bodies, disfigured by torture, sends a chilling message to villagers, one that paralyzes people with fear, sentences them to silence, and strains the limits of trust. Such killings have occurred with startling frequency in Colombia, as the conflict intensifies.

Further along, in another village, a peasant whom I will call Edgar Vargas described to me some of the violence that had touched him and his community directly. As we walked in the middle of his fumigated cornfield, Vargas recounted how, three years earlier, the army killed a day laborer employed by his son. The soldiers dressed the corpse like a guerrilla, hung it from a tree in a nearby field for several days as a warning to others, and then took the body away in a helicopter. More recently, the paramilitaries murdered his godson and gouged the boy's eyes out. Vargas added that he once owned a pickup truck, but the military took it from him, and he had recently seen the paramilitaries driving it. He now worried about his cattle, which he had moved to more distant pastures so that the paramilitaries, whom he described as people who "killed without compassion" (*mata sin compasión*), did not steal them too.

His story raised a number of questions. Who had fingered the laborer and Vargas's godson, and why had they done so? Did they live in the village, continuing to terrorize others? Did the dead men know their accusers, and if so, what was their relationship to them? Were the men FARC guerrillas or guerrilla collaborators? Or perhaps the murders were "mistakes" for which the perpetrators later offered compensation to the family?[27] Who committed the murders, and how could they ever live with themselves again? How was Vargas protecting himself against a similar fate? These and other questions ran through my mind, as Vargas's story evoked horrible images, but because of our circumstances, I could not ask them, for my own safety and his. That was, in part, how terror operated. It imposed a wall of silence between people, and fed on rumor, speculation, half-truths, and unspeakable certainties that were widely known. Speaking out was impossible, and because of widespread fear, many deaths were not reported.

This kind of terror disrupts local-level social relationships, and it enforces a sense of isolation. The survivors suffer behind a wall of silence without the emotional support of kin and neighbors, and an absence of trust and pervasive fear make it extremely difficult for terrorized peoples to construct the social ties necessary for any kind of collective action or mutual support. As they struggle to go on with their lives, people must constantly confront the physical reminders of terror that haunt their communities—massacre sites, trees from which victims were hung, places where loved ones were murdered, and unmarked graves. Because of the violence and these horrible memories, some peasants have fled over the border into Ecuador, or they have resettled in the anonymity of sprawling, urban squatter settlements that have mushroomed on the outskirts of Colombian cities as the number of internally displaced has grown. In Putumayo, over five hundred displaced families reside on the green hillsides around Mocoa. According to the parish priest, these families are headed primarily by widowed women who occupy makeshift dwellings with their daughters, who are often themselves widows. Many of the women suffer from severe emotional trauma, after witnessing the grisly deaths of spouses and sons, but there are no treatment programs for trauma survivors. The women find little relief. Paramilitaries virtually control the town of Mocoa and can continue to terrorize these women with complete impunity.[28]

The impunity-generated fear and insecurity that pervades the rural communities and urban settlements of displaced peoples is amplified by informers, who, either through direct coercion or the benefits that collusion offers, become incorporated by the armed actors and turned against others like themselves.[29] The creation of informers is a tactic that extends the repressive apparatus of the state—and other armed actors—into the fabric of daily life. In the context of Colombia's civil war, the tactic exposes individuals to retaliation by whatever armed group they happen to betray, but perhaps more importantly, it intensifies local-level violence in a range of different settings. Long-standing family feuds, political infighting, and interpersonal animosities can become deadly when these cleavages are exploited by army agents, paramilitary spies, and guerrillas. The perpetrators of violence in these circumstances may continue to live side-by-side with victims and survivors.[30]

As people are exposed to state- and paramilitary-sponsored violence, or simply left unprotected by the state in a context of growing instability, the privatization of security increases along with growing suspicion among and between different social groups. The presence of guerrillas and paramilitaries in rural Colombia is partially rooted in this process, and for peasants in Putumayo and other regions torn by violence, the formation of clientelistic relationships with the armed actors becomes part of local survival strategies. Whether these relationships are between peasants and guerrillas, drug lords, paramilitaries, or the army, they are hierarchical, authoritarian forms of social organization in which people find themselves receiving protection in exchange for information and material support. Abrupt shifts in the local balance of power, however, can cause defections from these networks, and this, in turn, heightens the sense of insecurity.

As the military situation changes over time, the "frontiers" between the various protagonists become the sites of intense and often indiscriminate violence (Pécaut 1999). For example, peasants who enter the small towns of Putumayo are often accused of being guerrilla collaborators, and urban dwellers who travel to the countryside are suspected as paramilitary spies. A local priest, who described the paramilitary takeover of one Putumayo town, said that a measure of predictability had returned to daily life, at least momentarily, now that it was no longer a zone of contention with the guerrillas. Some townsfolk felt slightly more secure as the threat of physical violence abated. The paramilitaries had imposed a strict "discipline" on local residents that was welcomed by some and feared by others. Although the extrajudicial executions and unpredictable outbursts of violence that accompanied the paramilitary arrival had ceased, townspeople were paying a high price for "peace," as they were obliged to operate within new channels of authoritarian clientelism, and community organizations were subjected to the dictates of the armed overlords or destroyed altogether. The paramilitaries instructed the priest to "just carry out his role," and they came to mass to see that he did. "They appear very Catholic," he remarked ironically. "You have to be very careful how you interpret the Bible."

In rural areas under guerrilla control, peasants related to the FARC and paid its tax with varying degrees of enthusiasm. In southern Colombia and Putumayo, where support for the FARC has historically been strongest in peasant frontier settlements, some people viewed the tax as a legitimate contribution and paid it willingly. Other people were either indifferent or hostile to the guerrillas whom they viewed as the latest group of demanding overlords. Sympathies fluctuated in accord with the guerrillas' ability to protect the local population and regulate the production and sale of coca on terms favorable to the peasantry. They were also influenced by the size of the tax demanded by particular commanders and its relation to the prevailing price of coca. Everyone understood that if they did not pay the guerrillas, they would be subjected to reprisals. They also knew that the paramilitaries and the army would extract a similar contribution on terms that were likely to be less favorable to rural people than those laid out by the guerrillas.[31]

Accepting protection from the guerrillas or the paramilitaries did not necessarily mean a political identification with these groups. As Pécaut notes: "A whole range of relationships can exist, from active participation . . . by militias . . . or open and covert civilian sympathizers, to straightforward sympathy and passive forms of acquiescence. . . . Even though all must submit to the constraints of their 'protectors,' many are able to see this relationship as principally an instrumental one. Individuals' accommodation strategies differ, but it is not unusual for them to undertake a rational calculation that takes into account both the advantages conferred by the armed groups' continued presence in the area, as well as the cost of their own obligatory obedience to such group's rules" (1999, 150).

The Creation of "Outlaws"

Impunity not only protects military and paramilitary perpetrators and allows them to continue terrorizing unarmed civilians. It is also simultaneously tied to the criminalization of ordinary people through the laws and policies enacted by government officials who are less accountable to their own citizens than to the dictates of the U.S. government and international financial institutions. In the Andes, impunity creates "outlaws" in the sense that it literally places people "outside the law." These people include recently criminalized sectors of the population, such as the Bolivian coca growers, and more generally, those who have been excluded from the benefits and protections of citizenship that form the basis of a democratic society. People turn to coca cultivation as a means of survival, because neoliberalism has impoverished them and destroyed other possibilities of livelihood. Poverty gives them few other options, and it polarizes society and degrades the poor in such a way that dominant groups can easily view themselves as superior. States then have less difficulty justifying the use of violence against criminalized groups, and the violence, in turn, reinforces their impoverishment and exclusion. The tightening vise of the drug war places rural people in Putumayo and the Chapare in an impossible position: abandoning coca cultivation means relinquishing their only hope of subsistence security, but by continuing to pursue an outlawed activity, they expose themselves and their families to the official violence of the state and, especially in the Colombian case, to the illegal violence of the armed actors who contend for control of the profits that flow from coca cultivation.

The experiences of Bolivian Rolando Quispe provide an entry point to understanding these processes. Quispe is a migrant from the highland department of Postosi, which he left in the midst of a severe drought that devastated the region in the 1980s. Although he is only forty-two years old, Quispe looks closer to sixty. His legs are thin and his shoulders are stooped. He is missing teeth, and his cheeks are sunken. His frail appearance hides a strong determination to defy the demands of the U.S. drug warriors and to continue growing coca, which he considers his right, but Law 1008 has taken that right away and made Quispe an outlaw. The law, and the impunity that makes its enactment possible, have also assaulted the moral universe in which he and other peasants live. What was right is now wrong, at least by the new standards established by Law 1008, and this shift in thinking is being felt in public schools.

One Sunday afternoon, Quispe and I met at a student fair in the settlement of Chipiriri, where his two young daughters attended school. A series of display booths lined a large outdoor patio and exhibited a variety of student projects in which coca and alternative development figured as prominent themes. They testified to the ideological pressure experienced by peasants to abandon coca. At one booth, a group of grade-school children had studied the banana, and they surrounded an exhibit that described the qualities and history of the fruit. A little girl handed out a questionnaire to visitors who were asked to respond true or false to a series of questions. One question asked, "Is the banana a marketable crop?" For grade-school students in Bolivian cities whose families regularly purchase bananas, the answer would perhaps be obvious; the question itself might even seem a little silly. But for Chapare children, the answer was less straightforward. Although their parents occasionally sold a stem of bananas to local middlemen, the fruit was typically planted in the vicinity of households and consumed as a snack. Yet these children were being encouraged to think of bananas as a cash crop that could provide an alternative to coca. When I circled "no," the girl corrected me, and when I asked her to explain, she replied that bananas were sold in markets. Her parents and other adults knew, however, that bananas were not a substitute for coca.

Replacing coca with bananas is in fact a peculiar proposition. All banana export zones are on coasts and easily accessible to ocean ports, and most world banana production is vertically integrated under the control of multinational corporations, which contract with small holders to produce the crop. In addition, bananas are a delicate fruit. They fall prey to numerous diseases and require careful packing procedures. Foreign consumers also place a premium on the size and appearance of the fruit. Not surprisingly, constant technological intervention is required during every phase of production and transportation. In the Chapare, there are no stable roads or packing facilities, and technological assistance through the alternative crop development program is rudimentary or nonexistent. To believe that peasants of a remote inland region could become competitive banana exporters is therefore a difficult stretch of the imagination.

Quispe and his family had managed to live in the Chapare for many years because coca, unlike other crops, provided them with a reliable cash income, but this all changed in 1999. Soldiers banged on their door in the early morning and announced that, because Rolando had not destroyed his coca bushes, they were going to do the job for him. A few of them forced the family to remain in the house, while others went to the field and set to work uprooting the bushes. They did the same thing that day to other families in the sindicato. Then, six months later, the soldiers returned and razed new coca plants that had sprouted up in the interim. I asked Rolando how he and his family were surviving. "Growing coca," he replied in a matter-of-fact manner that acknowledged no irony.

He explained that he could not return to the highlands, where his exhausted land produced very little, and migrating to the departmental capital of Cochabamba was not an option, because he felt that he was too old to look for work in a city already saturated with impoverished migrants. He also claimed that the U.S. government–sponsored alternative development program for the Chapare did not work for him. The crops promoted by local technicians—palm hearts, bananas, pineapples, coffee—could not be produced commercially on his land, which was hilly, rocky, and too far from a stable road to market the crops successfully. He therefore continued to plant coca but no longer in a single plot. He had adopted a new strategy of dispersing the bushes throughout the jungle in the hope that, if the soldiers returned, they would not find all of them, and he described the amazing ability of coca to grow just about anywhere; it would, he insisted, even grow out of rocks.

Quispe was not the only peasant around Chipiriri who refused to abandon coca cultivation, and peasant refusal raised the cost of repression for the state, as well as the danger faced by peasant families. Sporadic confrontations between peasants and the armed forces continued in Chipiriri and throughout the Chapare after the initial wave of forced eradications ended. Chapare peasants, lead by their charismatic spokesman, Evo Morales, tried continually to negotiate changes in the government's hardline "zero coca" policy, but state officials, pressured by the United States, refused to budge. A cycle of negotiations and confrontations continued, and, by early November 2001, tensions had once again reached a boiling point. The coca growers threatened to blockade the highway through the Chapare if the government rejected its proposal to allow every peasant family approximately an acre and a half of coca. The government, for its part, made clear that under no circumstances would it permit the disruption of transportation through the Chapare,[32] and the military sent four thousand soldiers to reinforce the eight thousand troops already stationed in the province. The minister of justice announced that human rights was not their first consideration. Once again, peasants found themselves in an impossible situation.

Colombian coca growers faced an even more desperate scenario. Violent clashes between the guerrillas, paramilitaries, and the army displaced some ten thousand people from their homes in Putumayo in 2001, and the aerial fumigation of coca fields aggravated the process of displacement. The toxic herbicide that rained down on fields posed stark options for peasant families. If they chose to replant, it could take at least ninety days for the soil to regain its fertility, but, with their food crops in ruins, they needed a cash income to purchase foodstuffs. Some young people opted to join the armed combatants as a means of survival. Other disillusioned peasants decided to set off deeper into the Amazon forest and carve out homesteads in less accessible regions where the spray planes might not find them. Yet replanting coca in a new location was no guarantee that the same thing would not happen again, and it contributed to the growing deforestation of Colombia's tropical forests, aggravating an environmental disaster impelled by the fumigation drive. Those who departed were not always able to sell their land and recuperate their investments, either. In such cases, they simply abandoned their holdings, which were then taken over by one or another group of armed actors and used to produce coca.

In the fields of peasants who had chosen to stay, bright green coca

Figure 13. The hardy coca bush resprouts after aerial fumigation. Photo by author.

plants had resprouted less then two months after a second round of fumigation had laid waste to the fields in San Miguel and Valle de Gua-muez. Peasants had pruned the poisoned bushes, and the roots had sent new offshoots to the surface as if by magic. The lush green leaves testified to the hardiness of the coca plants, which were probably more herbicide resistant than ever as fumigation acted as a form of selective breeding, and they stood in stark contrast to the bleak landscape surrounding them. More than at any time in the past, peasants depended on the budding green leaves for their economic well-being, and a state policy designed to destroy coca cultivation was having the opposite effect.

Impunity-generated political and economic violence wreaks havoc on the lives of impoverished people in Putumayo and the Chapare, as con-tending social groups negotiate space for themselves in a more globalized economy. On the one hand, the violence deployed by militaries, paramili-taries, and counternarcotics police forces is central to the consolidation of neoliberal capitalism. It deprives peasants of their most important source of subsistence, drives them from their lands, and destroys the social relationships through which they care for each other and channel collective demands for peace, economic security, and social well-being. All of this creates new, and aggravates old, inequalities. As ordinary people become more vulnerable, they are available for incorporation into, or exclusion from, the social relations of neoliberal capitalism on terms to which they have not agreed. Yet the social decomposition that arises from the state's own policies, which are themselves not always coherent, is never entirely controlled by the state, nor are newly marginalized people ever completely incorporated into new productive arrangements, as the failure of alternative development so vividly demonstrates. As coca growers continually reconfigure an array of social, economic, and political relationships, understandings, and forms of struggle amid the chaos imposed on them, they continually challenge the status quo. Order and disorder are thus closely connected in the Andes. The efforts of the United States and its local allies to impose order through the training of repressive security forces at the School of the Americas and by the Special Forces, draconian laws, and destructive economic policies is an inherently violent and contradictory process. It is a process that emerges from, and is structured by, impunity.

CHAPTER 8

Targeting the "School of Assassins"

Sunday, November 18, 2000, was the day chosen by anti-SOA organizers to commemorate the eleventh anniversary of the massacre of six Jesuit priests, their housekeeper, and her daughter by members of the Salvadoran security forces and to demonstrate against the School of the Americas. Twenty-one of the soldiers who planned, executed, and covered up the massacre had attended the School of the Americas, and the Atlacatl Battalion, to which they belonged, had received training at Fort Benning. In the aftermath of the murders, gruesome pictures of the dead victims lying in pools of blood flashed around the world and provoked an international outcry. The massacre intensified questions about U.S. policies that justified training security forces with abysmal human rights records, and it became emblematic of the Salvadoran military's brutality in a conflict that claimed thousands of civilian lives and lasted twelve years. The Jesuit murders had a profound impact on the core group of religious activists who organized to close the School of the Americas, and they became a potent symbol for the movement.

The weather could not have been more uncooperative for the demonstration outside the gates of Fort Benning. A continuous, bone-chilling drizzle fell from the leaden skies across southwest Georgia, and forecasters predicted more of the same for the day ahead. The temperature hovered in the upper thirties. Even with multiple layers of clothing, it was difficult to ward off the dampness and the chill, but the miserable conditions did not deter some ten thousand demonstrators who had come from as far away as Alaska to attend the annual protest. The protesters had come in cars, planes, and buses, and it was impossible to find a vacant hotel room in the entire city of Columbus.

Thousands of protesters crowded along the slippery edge of Fort Benning Road, preparing to witness a funeral procession that would memorialize the Jesuits and other victims of SOA graduates and culminate two days of anti-SOA protest. They huddled under umbrellas to remain dry and sipped coffee, blew on their fingers, and stamped soggy feet in an effort to stay warm. Hundreds of other demonstrators formed a long column that snaked down a side street just outside Fort Benning's main gate. They carried crosses, each bearing the name of a Latin American murdered by an SOA graduate. A few wore long black shrouds and white death masks. Others carried coffins. Together, they formed the heart of the mock funeral procession that would commit civil disobedience by marching illegally onto Fort Benning. As the procession prepared to get under way, the mourners fell silent. A speaker began to read out the names of the dead and the disappeared in a solemn tone over a public address system. After each name, a single drum beat sounded, and the crowd chanted *"presente."* The mourners marched four abreast into Fort Benning, where they risked arrest, prosecution, and possible prison sentences.

Protesters accused the School and the U.S. government of sponsoring terrorism and called for the end of taxpayer-financed military instruction for Latin American security forces at the SOA. Their dogged efforts over the last ten years had transformed a relatively obscure army school into a public pariah and pushed Congress to within a few votes of shutting down the institution. More broadly, their movement kept alive the memories of the victims of military atrocities in the Americas, even as the United States Army and allied militaries struggled mightily to obliterate them. As protesters brought the policies of the United States into sharper focus, Washington propagandists and military strategists had to confront fellow citizens demanding an end to U.S.-sponsored state terrorism in their name.

The movement did not spring forth overnight. It was the outgrowth of the tireless activities of a dedicated group of organizers and the moral outrage of thousands of people who felt compelled to speak out against the practices of their government. This chapter examines the 1990s campaign to close the School of the Americas. It considers the kind of people who joined the movement and asks why and how they mobilized to close the SOA at a time when the civil wars in Central America were winding down, the cold war was over, and Latin American countries had moved from military dictatorship to civilian rule. To answer these questions, my analysis first explores the core group of largely religious activists who initiated, organized, and led the movement. It argues that these people of faith—mostly middle-aged Christians—developed a long-term commitment to change through their involvement with previous social-justice causes, particularly the U.S. Central America solidarity movement of the 1980s. They understood the human consequences of U.S. policies in Latin America because, unlike most U.S. citizens, they had worked or traveled in the region, and they had met Central American refugees in the United States through their local congregations. They were also well positioned to disseminate information about the dark, unofficial side of U.S. government policies and activities in Central America through the religious organizations and faith-based communities to which they belonged. What they had learned offended their religious sensibilities and sparked similar feelings of moral outrage among other churchgoers in the United States who heard them speak. Righteous anger, however, was not enough to generate a grassroots movement that required a strategy and an organization.

To understand the emergence of the anti-SOA campaign, the discussion explores the development of an organizational structure that facilitated the dissemination of information, the creation of networks of solidarity, the involvement of young activists, and the coordination of resistance. By considering the broader relations of political power within and against which activists operated, it also examines how the movement dealt with the opening and closing of political opportunities that were largely beyond its control. Finally, the analysis focuses on the shifting tactics employed by the movement and the military to deal with each other and to portray a particular image of the SOA. The chapter concludes with an assessment of the anti-SOA campaign's successes and failures on the eve of the terrorist attacks against the United States on September 11, 2001.

Getting Started

In mid-November 1990, on the first anniversary of the Jesuit massacre, three men—Catholic priest Roy Bourgois, former priest Charlie Liteky, and Liteky's brother, Patrick—entered the School of the Americas with a letter to the commandant demanding that the School stop training Latin American militaries. The Liteky brothers placed the letter and pictures of the murdered Jesuits beneath photographs of the School's most distinguished graduates and members of its Hall of Fame. The men poured blood donated by activists over the pictures, the walls, and the floor and then rejoined Bourgois, who had just planted a cross in the ground at the entrance to the building. The three intruders spilled more blood to simulate a massacre and lay down to wait for the inevitable to happen. Before long sirens started to wail. Military policemen arrived and took them to a police station in downtown Columbus, where they were charged with trespassing and property destruction. All three were prosecuted and convicted. The Liteky brothers received nine-month prison sentences, but Bourgois, who had served jail time for a previous Fort Benning protest, got fourteen months.[1] Their action marked the beginning of the movement to close the School of the Americas; it constituted the first act of what became a tradition of nonviolent civil disobedience that grew to embrace thousands of U.S. citizens.

The protest also signaled the start of the School's gradual emergence from relative obscurity. Little known outside military circles, the institution operated for most of its life in the relative anonymity of Panama, far from the attention of critics and the prying eyes of the media, and it had quietly pursued its training mission at Fort Benning since 1984 with the support of a small group of influential civic boosters (see chapter 1). For the residents of Columbus who were aware of its presence, the School was merely an extension of Fort Benning, the economic powerhouse that generated jobs and contracts for members of the local community. Few objected to its activities, and almost nobody on the U.S. left—especially the secular left—knew of its existence. Not surprisingly, the movement to close the soa grew very slowly, and it initially received little attention.

Antiwar activism in the United States had waned by the late 1980s. The fall of the Berlin Wall in 1989 and the collapse of communist regimes in the Soviet Union and Eastern Europe brought the cold war to an end. Some military bases closed in the United States, and cuts in the defense budget heartened peace activists before the onset of the 1991 Gulf War racheted up defense spending again. Although Central America was still not at peace, the Sandinistas and the Contras signed a cease-fire accord in 1988, and the Salvadoran guerrillas agreed to a final peace agreement with the government two years later, ending the protracted civil war and placing Central America on the back burner of U.S. politics. All of this contributed to the demise of the U.S.-based Central America solidarity movement, but, as sociologist Christian Smith observed, the movement "educated and trained tens of thousands of U.S. citizens in the philosophy and methods of grassroots activism and disruptive political rebellion." He went on to suggest that by so doing, it might serve as a "critical wave in the long, historical tradition of grassroots peace and justice activism" (1996, 377).

Others students of social movements note that, although movements rise and fall, there are frequently important ties between historically distinctive eruptions of collective action. As Edelman notes, inspiration may "live on in the accumulated experience of . . . participants even after the movements themselves have faded away" (1999, 197). McAdam refers to "enduring activist subcultures" (1994, 43) that give continuity from one movement to another, and Hirschman discusses how the experiences of collective action may be preserved as latent "social energy" that becomes reactivated again in a different form (1988, 8). These phenomena are very common, and they describe how the Central America solidarity movement served as an important organizational and inspirational bridge to the contemporary campaign to close the SOA.

The core group of activists who initiated the anti-SOA movement were veterans of the 1980s struggles to end U.S. government support for murderous Central American regimes. They drew inspiration from the teachings of liberation theology and firsthand encounters with the victims of Central American political violence, and this inspiration lived on among them, even as Central America ceased to be a major political issue in the United States. The activists were middle-aged, church-affiliated, religious people. Many were (or had been) priests, nuns, and pastors who had worked as missionaries in Latin America, where they bore witness to the human consequences of the dirty wars, or they were involved with Central American refugees in the United States, primarily through the church-based Sanctuary movement. They understood that widespread impunity for military perpetrators was the norm in Latin America. A series of high-profile religious murders in El Salvador during the 1980s shocked and angered them: first, the assassination of Archbishop Oscar Romero, one day after he ordered the military to stop the repression; then, the rape and murder of four U.S. churchwomen; and ultimately, the 1989 massacre of the Jesuits. As more information about the School of the Americas became public, they were outraged that their government continued to spend U.S. tax dollars to train unaccountable security forces with miserable human rights records. When the Central America solidarity movement fell apart, taking on the SOA was an extension of the work that many had already begun. It flowed from their sense of moral outrage, the violation of their deeply held Christian beliefs, and a shared and enduring commitment to progressive social change.[2]

The irrepressible Roy Bourgois emerged as the movement's most articulate and charismatic leader. Upon his release from prison, he returned to a tiny apartment just outside the main gate of Fort Benning, and, with his neighbors looking on, dedicated himself to SOA Watch, the primary organizational structure that monitored the SOA and facilitated the mobilization and coordination of the campaign to close it. Born in Louisiana in 1938, Bourgois served in the navy for four years. He spent one year in Vietnam and returned to the United States deeply disturbed. His intention had been to make the army a career, but the war changed the way that he thought about the U.S. government and its armed forces. He left the military and entered the priesthood, and, in 1992, he was ordained in the Maryknoll order. The Maryknolls exposed him to the powerful influence of liberation theology, which taught that the church must make "a preferential option for the poor" and participate in the struggle for social justice on the side of the oppressed.

Liberation theology was shaped by neo-Marxist social analysis and emerged in Latin America during the 1960s, when revolutionary and reformist thinking was sweeping the continent in the aftermath of the Cuban revolution. It reached progressive religious groups in the United States in a number of ways. The Maryknoll publishing house, Orbis Books, translated several major works of liberation theology, including Peruvian Gustavo Gutiérrez's *The Theology of Liberation* (1973), which sold over one hundred thousand copies. This and other books were widely read by Christian seminary and divinity school students, and they were available in religious bookstores around the country. Missionaries and religious workers also disseminated liberation theology in the United States after direct exposure to it through contact with church people in Latin America, and, by the end of the 1970s, most U.S. Catholics and mainline Protestant churchgoers knew something about it (Smith 1996, 146).

Bourgois and the core group of anti-SOA activists absorbed liberation theology and strongly identified with its ideas about siding with the oppressed, political struggle, justice, and a God of the poor. Liberation theology was not an abstract religious doctrine for them. It resonated deeply with what they had seen and heard from ordinary Latin Americans who struggled to survive amid the political violence that surrounded them and who occasionally set out on dangerous journeys to the United States to escape persecution. And it continued to do so, even as conservative power brokers within the Catholic Church undermined liberation theology's institutional foothold by reassigning the progressive religious and speaking out against their practices.

Bourgois had pursued his calling in Bolivia, where he worked for five years in a poor neighborhood of La Paz during the repressive rule of General Banzer. The military viewed Bourgois with suspicion and expelled him from the country after he accused the regime of torturing prisoners. He returned to the United States to work in impoverished inner-city communities and became involved in the antinuclear movement, for which he produced a documentary film, *Gods of Metal*. After the murders of the U.S. churchwomen in El Salvador—two of whom were his friends—he began to speak out against U.S. involvement in Central America during the 1980s.

Former priest Charlie Liteky was another Vietnam veteran who came to the anti-SOA campaign from the Central American solidarity movement. Liteky devoted much of his time in the 1980s to demonstrating against Ronald Reagan's support for the Nicaraguan contras and the administration's backing of right-wing governments in the region. Many refugees whose relatives had been tortured and killed by U.S.-trained militaries fled to San Francisco, where Liteky lived. Although they were treated officially as illegal aliens subject to deportation, the development of the church-based Sanctuary movement provided protection for some, and Liteky's church was among those that voted to become a sanctuary.[3] Liteky began to attend evening meetings with refugees who were invited by parishioners to discuss their experiences. He and others sat in stunned silence as the immigrants recounted hair-raising stories of gruesome murders, narrow escapes, and pervasive terror. The tales shocked Liteky. They transformed the Central American conflict from an abstract issue broadcast in sound bites on the nightly news to a concrete reality embodied by traumatized human beings who continued to suffer from the emotional consequences of their experiences.

Liteky began to educate himself about Central America. He read *Witness to War*, a firsthand account of the Salvadoran conflict written by

Figure 14. Father Roy Bourgois leads anti-SOA protesters outside Fort Benning. Photo by author.

Figure 15. Former Salvadoran Defense Minister Vides Casanova (left) ordered the murder of three U.S. nuns and a lay worker in 1980 and was invited to speak at the SOA in 1985. Photo by Jeremy Bigwood.

fellow Vietnam veteran Charles Clement, who had provided medical care to the FMLN guerrillas. This book and the political analysis of Noam Chomsky exerted a major influence on Liteky's thinking and eventually prompted him to travel to Central America to see for himself. In 1986 he went to Nicaragua, Honduras, and El Salvador with a group of Vietnam veterans organized by Clements, and he subsequently returned to Nicaragua with a delegation sponsored by a faith-based organization opposed to U.S. aid to the Contras. He did not like what he saw and decided to leave his job as a counselor with the Veteran's Administration to dedicate himself to political activism.

Antimilitary protest was a new path for Liteky. The son of a navy noncommissioned officer, he grew up on or near military bases in Hawaii, California, and Virginia and felt comfortable in the presence of servicemen. He entered the priesthood in 1960, and six years later, in response to the army's call for chaplains, joined the military and set off for Vietnam, completely behind Lyndon Johnson's cold war crusade in Southeast Asia. In his opinion, the growing number of U.S. antiwar protesters were cowards, and unlike these un-American long-hairs, he intended to serve God and his country in Vietnam. He had an opportunity several months after arriving in Southeast Asia, when he was ordered to participate in a search-and-destroy operation with a company from the Fourth Infantry Battalion. The unit was surprised by the North Vietnamese and came under a heavy, unrelenting attack that pinned the Americans to the ground. As the dead and wounded piled up around him, Liteky dragged twenty-three soldiers to safety. For this act of bravery, he won the Congressional Medal of Honor, the United States' highest military honor and one which only some 3,400 people have ever received. President Johnson told Liteky during the award ceremony that he would "rather have one of these babies than be president."[4]

By 1986 none of this mattered to Liteky. His growing awareness of the human tragedy unfolding in Central America had recast the meaning of the Vietnam War, which seemed to be replaying in El Salvador, Guatemala, and Nicaragua. Disgusted with the Reagan administration, he renounced his medal and its lifetime, tax-free, $600-a-month pension. In a ceremony at the Vietnam Veteran's Memorial in Washington, D.C., Liteky left the award along with a letter addressed to President Reagan. He was one of only two recipients to ever give back the medal, and the first to do so for political reasons.

When I met the sixty-eight-year-old Liteky outside the SOA Watch office in November 1999, he wore blue jeans, a T-shirt with an anti-SOA slogan emblazoned across the front, and ratty tennis shoes. He seemed the antithesis of the carefully groomed, fastidious army officers of Fort Benning who inhabited a world within which Liteky had once moved easily. Tall and gangly with unruly gray hair, he showed me a number of mock coffins that he had built for the upcoming protest vigil. He was also preparing a memorial wall—inspired by the Vietnam Memorial—for the Latin American victims of the SOA graduates. He had left the priesthood several years earlier, unwilling to abide by its requisite celibacy, and was happily married to a former nun who had introduced him to Salvadoran refugees in San Francisco. Although he was still based in San Francisco, Liteky had been spending a lot of time in Columbus to provide a sustained SOA Watch presence as other leaders served jail terms. He rented a small house just up the block from the movement's cramped office, and every morning as military personnel entered Fort Benning, he held a silent prayer vigil for the victims of the SOA graduates in front of the base. His days in Columbus were numbered, however. Liteky was scheduled to go on trial for another act of civil disobedience against the SOA, and he expected to be prosecuted and sentenced to jail time for the action. This was, in fact, what happened. Seven months later, Liteky received a one-year sentence and a fine of ten thousand dollars.

Like Charlie Liteky and Roy Bourgois, the Reverend Carol Richardson's participation in the anti-SOA movement emerged from her prior work against U.S. policies in Central America. And she, too, spent time in prison. She had pastored a church in Baltimore and traveled frequently to Central America in the 1980s. When Bourgois began calling attention to the SOA, she was working as the national grassroots coordinator for Witness for Peace, a faith-based organization that played a key role in opposing U.S. government support for the Nicaraguan Contras. Witness for Peace had organized delegations of U.S. citizens who were willing to sustain considerable risk by traveling to the Nicaraguan war zones to meet with the victims of U.S. policies and to hold highly visible peace vigils in full view of the Contras. The organization encouraged delegates to return home and speak to the media, congressional representatives, friends, colleagues, and church congregations about what they saw. Taking U.S. citizens to a war zone was a tactic that brought people face-to-face with the human suffering caused by their government, and most delegates returned to the United States enraged and ready to take action.[5]

It should come as no surprise that years later, many former Witness for Peace delegates found their way into the movement to close the SOA.[6] Richardson explained that in the early 1990s "almost immediately after I came on the staff of Witness for Peace, I started getting calls from the grassroots, saying that we really want to work on this School of the Americas issue. So it was from these calls that I was prompted to call Roy [Bourgois], who had been [working against the School] for about two and a half years." Richardson moved Witness for Peace into the anti-SOA struggle, but it would be some time before the movement grew into a force that threatened the School's existence and focused the energy and attention of its officials.

Demonstrations were typically small. Ten people recruited by Roy Bourgois from his personal network of social justice activists took part in a water-only fast in 1990, and the primary act of collective protest—a vigil in front of Fort Benning on the anniversary of the Jesuit massacre—attracted only a few dozen people in the early years. But by 1995 this situation began to change when, during the November vigil, thirteen people decided to risk arrest by marching across a white line that demarcated the boundary

Figure 16. Anti-SOA protesters organized by Witness for Peace demonstrate outside Fort Benning. Photo by author.

of Fort Benning. The line crossers were prosecuted and sentenced to jail, but they drew attention to the incipient movement and started a tradition that characterized future demonstrations. The number of protesters and line crossers increased in subsequent years, as more young people, veterans, and labor activists joined the struggle; in 1999, for example, twelve thousand people attended the November demonstration, and five thousand committed civil disobedience by trespassing onto Fort Benning. The experience of crossing the line energized activists by demonstrating their collective resolve to the army and strengthening their ties to each other and the movement.

Why, we might ask at this point, did the movement take off when it did? How did it develop from a sense of moral outrage among a small group of veteran activists, and how did it build an organizational structure capable of spreading information, connecting disparate groups of activists, recruiting new sympathizers, and making its demands felt among Washington policymakers? To answer these questions, we need to turn to the step-by-step nature of political organizing and the broader social and political forces that shaped the growth of the anti-SOA movement.

Building a Movement

Anti-SOA activists drew on the support of numerous congregations, faith-based communities, and religious organizations that loaned their facilities, contributed funding, and distributed information. This solidarity enabled movement organizers to reach a large number of people whom they would have had little ability to reach without the institutional support of this loosely defined, progressive religious community. The backing was crucial to building an organizational structure capable of sustaining a long struggle against the army and the government's efforts to preserve the School. Faith-based organizations, as well as colleges, universities, and secondary schools with religious affiliations, served as platforms from which activists spoke out against the continued U.S. training of Latin American militaries. The religious activists—priests, nuns, and pastors—who constituted the core group of the anti-SOA resistance were respected in these settings. What they had to say was taken seriously by those who heard their stories—primarily white, middle-class, and well-educated people of faith. As these people learned about the SOA, they began turning out for the annual protests and writing letters to congressional representatives, demanding the closure of the SOA. They shared a sense of entitlement that sprang from their life circumstances, and they expected the government to listen and to heed their demands.

Roy Bourgois was the most dedicated and tireless campaigner against the School of the Americas. Every year, beginning in the early fall, Bourgois left his tiny apartment in Georgia and traveled around the country, visiting churches, colleges, and high schools to speak about the School of the Americas. He encouraged his listeners to write to their congressional representatives, urged them to come to the November demonstrations, and invited them to Washington, D.C., for annual spring protests on the steps of the Capitol. He was a passionate speaker who presented the case against the SOA through examples of fallen religious leaders and discussions of peasant massacres, such as the one in El Mozote, El Salvador. He then listed the names of SOA graduates who ordered, participated in, or covered up these atrocities. Bourgois also moved his listeners with personal stories about Latin America and his experiences of protest and imprisonment.

Although these speaking engagements, which lasted well into the spring, were particularly effective ways of reaching people and communicating the reasons for closing the SOA, they were not the only organizing technique available to the movement. The use of home computers expanded throughout the decade, and e-mail, listservs, news groups, and the expanding World Wide Web became valuable tools for staying connected and informed. The Internet allowed national organizers to connect with small groups of disparate grassroots sympathizers and coordinate joint actions, and it facilitated the creation of networks of loosely aligned coalitions of activist organizations.[7] The Web also enabled organizers to post information that was either not published by the mainstream media or downplayed by it.

For example, after SOA Watch acquired a list of all the SOA's Latin American graduates through the Freedom of Information Act, it published the list on its Web page. The list contained each soldier's name, rank, and nationality as well as titles of courses taken and the dates of attendance. Subsequently, whenever a human rights organization or a truth commission published reports containing the names of military officers involved in human rights violations, activists correlated these names with those on the list of graduates. The results were shocking. SOA Watch not only linked alumni to some of the worst atrocities of the Latin American dirty wars; it also connected them to crimes that continued to occur long after the signing of peace accords in Central America and the replacement of military dictators with civilian rulers, crimes such as the murder of Guatemalan Bishop Juan Gerardi in 1999.

The growth of the movement also depended on activists' ability to capitalize on tactical mistakes made by the army and government, as well as on key moments of SOA vulnerability that the organizers helped to create but that they could not completely control. An important conjuncture, for example, developed in 1995, the first year that protesters crossed the line onto Fort Benning during the November protest. Thirteen individuals—including a Catholic nun—were promptly arrested after they set foot on the base, because the government claimed that their actions violated a court-mandated ban on political speeches and protests on military bases. Government prosecutors hauled the demonstrators before Federal Judge J. Robert Elliot, an old-line segregationist who had issued injunctions against Martin Luther King Jr. and had set aside Lieutenant William Calley's conviction for the murder of Vietnamese villagers in My Lai. Elliot took a dim view of the "SOA 13" and sentenced them to jail terms that ranged from three to six months, but his ruling was a tactical misstep that exposed the heavy-handedness of the army and the judiciary. It stoked the moral outrage of movement veterans and advanced the cause of the movement in a way that neither the judge nor the army anticipated.

The Leadership Conference of Women Religious—an organization that represents more than 78,000 Catholic nuns in the United States—was incensed because one of its members, Sister Claire O'Mara, received a six-month sentence. The organization passed a resolution to close the SOA and gathered over 25,000 signatures in support of a House bill to cut off the School's funding. Others in the religious community were equally outraged that a nun and other nonviolent resisters were imprisoned simply for expressing their views on an open military base. Public opposition to the SOA increased. When the thirteen prisoners of conscience got out of jail, prison seemed only to have increased their moral authority, as they continued to speak out against the School of the Americas and the system that put them behind bars.

Even more than the prisoners of conscience, however, the public revelation in 1996 of training manuals used at the School of the Americas sparked a dramatic upsurge in the movement. The six manuals, which were used at the School between 1982 and 1991, came to light as the result of a secret 1992 Defense Department investigation that surfaced in 1996, when Congress initiated a probe into the CIA's activities in Guatemala. The manuals blurred the distinction between armed guerrilla insurgents and unarmed, peaceful protesters, and they made no reference to the law when instructing Latin American militaries to spy on and infiltrate political parties, unions, and community organizations, as well as to "neutralize" them. The manual on "Terrorism and the Urban Guerrilla," for example, stated that "another function of the CI [counterintelligence] agents is recommending CI targets for neutralizing. The CI targets can include personalities, installations, organizations, documents and materials . . . the personality targets prove to be valuable sources of intelligence. Some examples of these targets are government officials, political leaders, and members of the infrastructure."[8] Although they were compiled from training materials used by the United States military in programs around the world, the "torture manuals" were linked publicly only to the School of the Americas. Editorials in major newspapers—the *Atlanta Constitution*, the *Boston Globe*, the *Los Angeles Times*, the *Chicago Tribune*, the *New York Times*, the *Cleveland Plain Dealer*, and other dailies—

called for closing the institution and briefly focused national attention on the School. The uproar spurred the formation of a bipartisan coalition in the House of Representatives that opposed the School, and the initiative begun by an intrepid priest and a handful of veterans from the Central America solidarity movement blossomed into a dynamic force that became increasingly difficult to ignore.

Record crowds began turning out for the November demonstration. The media started to take more interest in the event, and the army increasingly found itself on the defensive in an escalating public relations war. When seven thousand people crowded the Columbus side of Fort Benning in 1998 and more than 2,300 of them defied police orders and marched onto the base, the military surprised everyone by making no arrests. Deciding *not* to arrest marchers was a tactical move intended to trivialize the civil disobedience, undercut the significance of the protest,[9] and recapture the initiative for the army. Military policemen herded marchers onto waiting buses and then dumped them in a park, about a mile from the protest site. Demonstrators received letters that banned them from the post. The practice of giving so-called ban-and-bar letters continued, and for the next two years, only repeat offenders were prosecuted. One frustrated SOA official, who seemed to sense the futility of jailing demonstrators, complained bitterly in 1999 about an elderly nun who was scheduled to be released and who would, he remarked sarcastically, "go on the lecture circuit" with her stories. Although the military had little difficulty managing men with guns, dealing with angry nuns was apparently another matter.

With the growth of the movement, Carol Richardson opened an SOA Watch office in Washington in 1996, after staffing the Georgia office for six months while Bourgois was in prison. The Washington base facilitated the movement's legislative strategy on Capitol Hill and helped to coordinate collective actions in the national capital. Six regional offices also began operating around the country by the end of the decade. Coalition-building and organizing initiatives focused on colleges, churches, human rights organizations, and labor unions. SOA Watch produced a series of videos and informational materials that activists used when they gave talks. Organizers also held fasts; they set up meetings with congressional representatives and organized delegations to meet with them; they sponsored fact-finding trips to the Colombian war zones with Witness for Peace; and they occasionally staged highly visible protest action to call

Figure 17. Witness for Peace delegate and SOA advisory board member Ken Little protests U.S. policy in Colombia. Photo by author.

attention to their cause. At the 2000 Republican convention, for example, demonstrators reenacted a massacre perpetrated by School graduates, and during several days of national antiwar mobilization in April 2002, forty anti-SOA protesters disrupted the morning rush hour in Washington by riding their bicycles against the flow of traffic.

As the movement grew beyond the original group of middle-aged stalwarts, an SOA Watch advisory group was created in 1998 to coordinate strategy, incorporate the views of newcomers in the decision-making process, and continue reaching out to a broader constituency. Although religious leaders from the Central America peace movement retained a major presence on the board, carpenter Ken Little represented organized labor, Latin American history major Jackie Downing from Oberlin College spoke for the growing numbers of students, and Colombian Cecilia Zarate-Laun, the cofounder and program director of the Colombia Support Network, brought a key perspective on the militarization of the

Colombian drug war and the intensifying human rights debacle in that country. The growing strength of the movement became evident in July 1999, when the House of Representatives voted to cut approximately two million dollars from the SOA budget, an act that, had it been upheld by the Senate, would have effectively abolished the School.

The young people who started to flock to the SOA Watch–sponsored vigils in Washington and Fort Benning injected new creativity into the events by constructing large, colorful puppets, participating in "die-ins," and engaging in other acts of street theater. A larger number were students, who had learned about the School of the Americas through talks given by Father Bourgois at their colleges and universities. After attending the annual vigils, the students returned to campus, organized presentations, and drew others to the cause. Many of the young recruits were born in the late 1970s and 1980s, when resurgent conservatives and right-wing Christian fundamentalists sought to erase the progressive legacy of 1960s political activism, and the anti-SOA movement was their first experience of collective action. It exposed them to new ways of thinking that directly challenged some of their basic assumptions about politics and U.S. foreign policy and brought them together with an older generation of seasoned activists.

Not all the young people, however, were newcomers to collective action. Some had participated in anti–sweat shop campaigns on their campuses and an increasingly large number were active in the global justice movement. Those recruited from secular backgrounds were often surprised by the religious symbolism and the deep-seated faith and moved by long-time anti-SOA activists. Leone Reinbold, for example, learned about SOA Watch in 1998, when she was a student at Evergreen State College and was involved with global justice issues. She attended a presentation given by a friend who had demonstrated against the School the previous year, and she first participated in the November protest vigil in 2000. Like many young activists, the twenty-year-old Reinbold did not identify with the faith-based nature of the anti-SOA movement, but she was impressed with the way that SOA Watch used religion to organize collective action and to mobilize people who would not have felt comfortable in a more secular movement. She was also inspired by the older activists, whose experience and commitment set an important example for her. "It's really phenomenal to learn from people who have been organizing for thirty years," she remarked. "And it's interesting to be around people who say that six months in jail is no big deal. They give me a lot of hope to sustain my beliefs. It takes a lot of energy to just live."

Tacoma resident Bruce Triggs learned about the School of the Americas through a priest in the Catholic Worker movement. He participated in a 1994 fast on the steps of the Capitol when he was twenty-six and then went to the Georgia vigil and demonstration in 1996. Triggs and his wife played a key part in introducing puppets to the November protest. They constructed a giant puppet that they named "Our Lady of Guadelupe" and took it to the 1998 march. "The nuns loved it," he remarked. "They all wanted to have their pictures taken with Our Lady." He then met with other puppeteers at various demonstrations around the country and observed how global justice activists used giant puppets to transform protests "into more interesting marches that were not just people shouting at buildings."[10] During the "A16" protests in Washington, D.C., where global justice activists staged protests during the spring 2000 meeting of the International Monetary Fund, Triggs handed out leaflets to other puppeteers and urged them to come to a "puppet convergence" in Georgia before the fall vigil. He managed to recruit some fifty puppeteers, and for several days before the demonstration, they built large puppets that captured the attention of the media when they bobbed onto Fort Benning.

The growth of the annual Georgia protest had a significant impact on the Columbus economy that did not pass unnoticed by the military. For a long weekend every November, throngs of demonstrators filled the town's hotels and bed-and-breakfast inns, and they even spilled over into neighboring towns. They also patronized restaurants and fast-food establishments, and some rented cars. Many participated in nonviolence training at the downtown Bradley Theater, which SOA Watch rented. Columbus entrepreneurs began to relish the arrival of the movement. Hotel and inn operators practically stumbled over each other to court people with reduced room rates, and Atlanta-based Delta Airlines offered protesters 10 percent discounts on round-trip tickets. Their behavior was not a case of capitalism gone mad. The Columbus Chamber of Commerce staunchly supported the SOA and even passed a resolution that expressed this sentiment, but many of its members were simply not willing to pass up a golden money-making opportunity. An excited chamber of commerce representative articulated the enthusiasm of others when he gushed to a National Public Radio interviewer in 1999 that the anti-SOA

Figure 18. Anti-SOA demonstrators stage a "die-in" at the entrance to Fort Benning. Photo by author.

Figure 19. Giant puppets send a message to the military. Photo by author.

protest was even bigger than a recent Baptist convention, and business-men, he insisted, were going after the protesters "like any other conven-tion group."[11] Such statements ruffled the feathers of Fort Benning offi-cials, who felt betrayed by their fickle allies.

Yet despite the rapid growth of the movement after 1996, the anti-SOA campaign garnered little support from local people. Columbus was a mil-itary town in which many residents depended on Fort Benning for jobs. This was particularly the case in the African American neighborhoods that abutted Fort Benning on Columbus's south side. Small homeowners and apartment-complex dwellers served in the army and held low-level civilian positions on Fort Benning that participation in the movement would jeopardize. The leaders of their primarily Baptist churches did not have the same connections to Latin America enjoyed by their progressive, white counterparts in the mainline Protestant and Catholic churches. Latin American political violence had therefore not become an issue of concern for these people. In addition, poor and working-class people did not enjoy the same level of Internet access as the middle-class protesters who formed the bulk of the anti-SOA campaign, and as a result, their access to alternative sources of news and analysis was more limited. Fi-nally, even though the anti-SOA campaign drew heavily on the symbolism and tactics of the earlier civil rights movement, the primarily white, middle-class protesters who flocked to Columbus for the annual demon-stration had few connections to local African Americans.

Fifteen-year old Terence Jones and his fourteen-year-old cousin Kevin articulated to me some of what this distance meant on the day of the vigil. The boys lived in Southgate, the apartment complex just outside the main entrance to Fort Benning, and they were clearly animated by all the excitement as they watched protesters gather for the 2000 demonstration, but they were not clear about the religious beliefs that motivated so many mostly white people to come to the demonstration and that differed from those espoused by the military. When I asked them what they thought about the event, Terrence replied, "It's a good thing. They [the soldiers] killed people at a Catholic School. That's wrong. They did a sin, and now they are trying to be with God, too, or whoever ya'll worship."

The adults in the neighborhood were not participating in the march, but some of them were taking advantage of the opportunity to make a little money selling food and hot beverages to the protesters. Marion Robertson, a retired appliance technician, was feeding hamburgers, ribs, barbecue, and coffee to hungry demonstrators outside his ground-floor apartment. He didn't think that the SOA should train Latin Americans because the soldiers come from "Third World countries that are not going to be any real threat to us. So why waste the money training them and waste all our tax money?" His comments reflected those of others who felt that the training dollars could be better spent domestically.

As the protest campaign caused the army more grief and the congressional vote aggravated the internal hand-wringing, officials could no longer discount the growing opposition. Former commandant Roy Trumble stated that the movement "was and continues to be the major threat against the School," and he, like his successor, described his biggest challenge as contending with it. "SOA Watch was already active when I got here in '95," he explained, "and of course they have grown every year. They have had a fairly significant evolution. The core element is the same, but their ability to reach thousands of others and to manifest that in a protest movement is fairly significant. It was my number one challenge— how to deal with the issues brought about by SOA Watch. Find out what they were, address them, and see what we in the institution could do to remedy the situation, if it was an institutional thing, or answer the attacks." Trumble's depiction of the movement as a "threat"—an active, malevolent force—that launched "attacks" against a seemingly passive school spoke volumes about the army's view of SOA Watch and itself. No longer able to pretend publicly that the movement was of little concern, the army shifted tactics and sought out new ways to discredit the movement through spin control, public relations efforts, and harassment.

Of Marxists and Maryknollers

Retired general and SOA Board of Visitors member Paul Gorman chuckled as he recalled a meeting with a high-level official of the Ecuadorean government in the early 1980s. Gorman was on his first trip to Ecuador after assuming the leadership of the U.S. Southern Command. When he inquired about the political situation in Ecuador, the official replied that every morning he knelt by his bed and prayed to God to deliver the country from Marxists and Maryknollers. "I get the picture," responded Gorman with a sardonic grin. Less than twenty years later, Gorman and SOA officials might have been voicing a similar prayer, as Bourgois and ever larger contingents of protesters kept coming back to

Fort Benning. For some military officials, clear-cut distinctions between Father Bourgois, the Maryknoll order that supported him, and leftist Latin American guerrillas did not exist. More than one SOA officer described to me how Bourgois had traveled to El Salvador in the 1980s with a film crew and "disappeared"—a word that they spat out with contempt. They asserted that he joined the guerrillas and then, after "patrolling with the FMLN," resurfaced, claiming to have made contact with "the poor." Their message was clear: Bourgois was a guerrilla—or at the very least, a guerrilla sympathizer, a characterization that differed little from the ways that U.S.-supported Latin American militaries conceptualized domestic critics in their countries.

Whatever their personal thoughts about Bourgois, Pentagon and army officials understood that the movement was a force with which they had to reckon. In addition to prison sentences and fines, activists experienced various forms of harassment. On one occasion, for example, someone entering Fort Benning threw a cannister of tear gas at protesters fasting in front of the base, and Bourgois periodically received threatening letters and unsigned death threats, the frequency of which varied in accord with movement activities and broader political developments. His time and energy were also sapped by bureaucratic obstacles created by local officials intent on impeding the November vigil. Prior to the 2000 protest, for example, the Columbus city manager—a retired general widely believed to have more power than the mayor—refused to authorize the installation of a pole for the electricity needed to power a public address system, and only after Bourgois hired a lawyer and made calls to the media did the official finally back down. The army also harassed protesters by videotaping the November vigil and forcing those who marched onto Fort Benning to pass directly beneath a cameraman suspended above them in a cherry picker. All these acts were clearly designed to intimidate anti-SOA activists. Less clear were the organizational arrangements, if any, that linked these attempts to unnerve the movement. The intimidation was mild enough that most people were not sufficiently fazed to turn away from the cause, and it stoked a sense of anger that fueled their desire to close the SOA.

Beginning in 1996, the year the torture manuals came to light, SOA officials and their handlers in the Defense Department faced an enormous public relations crisis that the movement aggravated whenever possible. Worried about the future of the School, they began to refashion the curriculum, eliminating some controversial courses, renaming others, and designing a new group of feel-good classes to undercut a rising chorus of accusations that the institution was a training ground for dictators and assassins. "Psychological Operations" was axed, because, according to Colonel Weidner, it conjured up images of Chinese water torture, even though it was just about "advertising." It reemerged as "Information Operations," however. A sniper course was canceled, although students could still get training at the Fort Benning sniper school nestled in a secluded, pine-covered corner of the base just down the road from the SOA, where few critics paid attention to trainees practicing kill shots from four hundred meters. A series of new-look courses, such as Peace Operations, Democratic Sustainment, Humanitarian De-Mining, and Human Rights Train-the-Trainer, were also offered to burnish the School's tarnished image and highlight a new role for the institution in the post–cold war era (see chapter 2).

Fiddling with the curriculum was not the only new tactic. SOA officials also began to debate activists at events organized at colleges and universities. One such encounter at Georgetown University in 1999 featured the SOA's Lieutenant Colonel Homer Harkin, head of the School's Department of Joint and Combined Arms, and SOA Watch activist and University of St. Thomas professor Jack Nelson-Pallmeyer, author of the Obis book *School of Assassins*. Harkin began by announcing that he would not debate, because that was a matter for civilians. He would, however, provide information about the SOA so that members of the audience could make their own decisions. Although it was unclear whether the ensuing exchange changed anybody's mind, Harkin demonstrated that he needed to hone his skills as a public relations representative. Showing a remarkable obtuseness about social life on a college campus (even the relatively conservative Georgetown), Harkin came dressed in his military uniform accessoried with a variety of pins, patches, ribbons, and other paraphernalia and stated that he represented the sixth generation of military men in his family. The gap that separated him from the majority of college students could not have been more apparent. Then, apparently unaware of the university's fine Latin American studies program, Harkin displayed his ignorance of the region by making sweeping generalizations about the allegedly "communistic" nature of regional social movements that, unbeknown to him, represented diverse political perspectives.

Taking the battle to the movement on still another front, the Defense Department's Training and Doctrine Command (TRADOC) created a board of visitors in 1996 to provide more institutional accountability and to incorporate civilian perspectives in the SOA's curriculum. Board members were instructed to periodically review the SOA's courses and provide advice on School policy. Yet this effort to create greater transparency was just another public relations ploy. When the board convened for the seventh time in February 2000, it discussed some curricular matters but spent most of the three-day meeting pondering how to bolster the SOA's public image, get additional funding for the School, market the institution more effectively in Latin America, and discredit Roy Bourgois and SOA Watch. The eight-member board consisted of two retired ambassadors: chairman and former U.S. ambassador to the Organization of American States Luigi Einaudi and retired foreign service veteran David Passage;[12] two academics: American University dean Louis Goodman and Boston University political scientist David Scott Palmer, two military representatives: SOA commandant Colonel Glen Weidner and retired general and former SOUTHCOM chief Paul Gorman; one lawyer: Patton Boggs attorney Steven Schneebaum; and USAID functionary Johanna Mendelson Forman, the lone woman. For the first day of the board's deliberations, several generals and their entourages also participated. They included the TRADOC commander, General John Abrams; Under Secretary of Defense for International Affairs, Jack Speedy; the commander-in-chief of the U.S. Southern Command, Brigadier General James Soligan; and the commanding general of Fort Benning, Major General James LeMoyne.[13]

The proceedings began when the chair, Luigi Einaudi, posed the following question to the group: "Where does USARSA [United States Army School of the Americas][14] fit into the priorities of the system?" The question was followed by presentations from two generals who advanced the usual boilerplate nostrums about promoting democracy, stability, free-market economics, and prosperity throughout the Western Hemisphere. Einaudi then told the group, "We are at a critical moment. The critics of the School are taking center stage, and they are already saying that we are doing nothing new here." He instructed the members to discuss refounding the School.

Following Einaudi, Defense Department representative Jack Speedy took the floor and detonated a bomb blast that reverberated throughout the group for the next three days. The Secretary of the Army, he announced, had already drafted legislation to remake and rename the School, and it had been road-testing the proposal with the coalition of 230

House representatives who voted against the School in 1999. The strategy behind the legislation had two major objectives: to quell the annual protests against the SOA by dividing the movement, and to split the congressional coalition so that funding for the School would not be threatened in the future. The so-called changes laid out in Speedy's game plan were extremely minor and defensive in character. They included more tweaking of the curriculum, filling some teaching positions with civilian professors, and renaming the institution. The name change emerged from conversations with members of the House of Representatives coalition that had voted against the SOA. The Secretary of the Army's staff approached these representatives individually and asked each one what they needed to support the School. The consensus, according to Speedy, was that anything called the "School of the Americas" was too controversial. A new name, however, would give them the political cover necessary to support the institution. Other proposed innovations, such as giving military training to policemen and providing human rights instruction, were not new, but preserving the SOA *did* require something new: a public relations offensive that presented the "new" School as an effective break with the past, admitted no guilt, and placated domestic critics without alienating Latin American commanders.

By the end of Speedy's presentation, board members could not conceal their anger. Outraged less by the proposed cosmetic changes to the SOA than by the realization that they had been cut out of the policy-making process, members subjected Speedy to a withering cross-examination. The churlish General Gorman began. "I am unwilling to sit on a board of visitors like some kind of a potted palm for a political appointee in Washington," he fumed. Yet unlike other board members, Gorman was also upset by the content of the legislation, which, he felt, moved the School away from its core mission of providing military training to the allied armies of Latin America. "Diddling with the curriculum and putting in hokey courses," he charged, was simply unacceptable, and the name change was little more than "smoke and mirrors." Gorman later complained that he made a mistake in supporting the School's move to Fort Benning in 1984. If the institution had remained in Latin America, he insisted, the "problem [i.e., the protest] would not exist, because Americans were the ones causing it."

When it was his turn again, Einaudi stated that "the real damage that he [Father Bourgois] is doing is with the know-nothings in Congress who

want to support his off-the-wall views." He then told an anecdote about
some of the Latin American instructors who had mingled with the
marchers during the last protest vigil and who were impressed by the
restraint exercised by the police. He joked that the School should invite
Bourgois to teach a course on democracy. In a more serious vein he
allowed that the protests represented a real problem, but "we can deal
with it," he asserted, looking at Speedy.

After all of the verbal chest pounding and vituperation died down,
Einaudi brought the morning session to a close, and the members adjourned for lunch, mumbling discontentedly about "the process" and
threatening to place calls to policy insiders in Washington. When the
afternoon session convened, Einaudi informed members that various SOA
officials would update them on the present status and workings of the
School. The board then received briefings about curricular changes, the
shifting student body, staffing concerns, and the budget. The briefers
emphasized the need for an increase in funding for the SOA and a more
aggressive marketing campaign to boost enrollment that had diminished
due to rising costs and in-country training provided by Special Forces
Mobil Training Teams (MTTS). Board members concurred. The SOA's in-
imitable public affairs officer, Nicholas Britto, then gave his presentation,
which prompted considerable discussion about the best way to undermine the anti-SOA movement.

Britto began with video news clips from the local Fox television station
that portrayed the SOA in a positive light, and he continued with a slide
show. Three slides presented portions of the SOA Watch Web page, and
two of them laid out the spring travel plans and speaking tour of Father
Bourgois. As board members pondered these slides, General Le Moyne
noted that Bourgois intended to speak at American University in April and
looked to Dean Louis Goodman for some kind of a statement. Goodman,
however, had recently had a run-in with students angered by his participation on the SOA board, and he sidestepped the issue of engaging Bourgois
in a debate. "I wonder how it [Bourgois's talk] will be set up," he mused,
and then expressed doubt about taking on Bourgois "if [the exchange of
views] were broadcast to a large audience." "Caution and colorlessness,"
chimed in Einaudi, take symbols away from the protesters who might use
them against the School. "I think that is what Lou is saying." Goodman
did not demur.

USAID representative Mendelson-Forman added her two cents' worth of commentary. What always hits the U.S. public are pocketbook issues, she argued. Turning to Britto, she then told him that making more of an issue about the SOA Watch budget could be an effective tactic. Gorman piped up that a good public relations campaign could also paint Bourgois as un-American. He added that finding an academically reputable group to write favorable articles about the School was a useful tactic. Such articles could compare Father Bourgois to other supposedly subversive groups. Weidner opined that calling public attention to some of the statements made by movement leaders could highlight their extremism and isolate them from other sectors of the movement that were more sympathetic to the military. He cautioned that he did not want the School to be accused of monitoring civilian groups. Goodman suggested that members limit themselves to "shedding light" on matters.[15] The discussion continued in this manner until the group adjourned for the day without settling on any particular plan of action.

The board reconvened the next morning to discuss the specifics of the proposed legislation to reform the School and to start developing an outline for their final report. The size of the group had dwindled. Speedy and the generals were gone, much to the apparent relief of some members, but the remaining participants still smarted from the wounds inflicted the previous day by higher authorities in the Defense Department. Nevertheless, the atmosphere was relaxed, and board members settled in for a long day of work.

The group revisited themes from earlier sessions—the name change, the incorporation of civilians, and marketing the School. Some individuals worried that involving too many civilian instructors and students might cause morale problems in the military and generate a decrease in enrollment. Latin American commanders, they noted, did not want to spend valuable training dollars on human rights lectures delivered by Yankee civilians, and too much emphasis on human rights might also have a negative effect on enrollment. It became clear that marketing the School to foreign and domestic audiences required different strategies.

At one point, the issue of meeting minutes arose, and board members launched into a discussion about how to present them that bordered on the surreal. They had come to view the gathering as a closed-door, private event, even though the meeting was open to the public and had been advertised in advance. This was, perhaps, because I was the only member of the public present, and in the absence of greater civic participation, the public nature of the event seemed to have been forgotten. The discussion about minutes focused on how to conceal the identity of members in the official document and sanitize the nature of their conversations. Chairman Einaudi commented that after the last board of visitors meeting, he had spent considerable time "polishing" the minutes so that particular individuals would not look bad. Goodman then suggested that the minutes focus on the issues, rather than the individuals who expressed certain views, as a way to make them more usable to the board and to avoid controversy. David Passage agreed. The public, he felt, did not have to know where each member stood on the more controversial issues. The rest of the group concurred, although they recognized that minutes focused on issues required more work to produce than a simple rendition of who said what to whom. When reference to individuals was unavoidable, the terms "presenter" and "board member" were to be used, and the document that they produced would be notable less for what it said than for what it left out.[16]

On the last day of the meeting, Nicholas Britto passed out copies of an article from the morning paper in which Roy Bourgois discussed the recent assassination of Guatemalan Bishop Juan Gerardi and links to SOA graduates. Most board members had read it by the time I arrived, and they were venting their irritation. Referring to Bourgois—and not the Guatemalan security forces—Einaudi ranted, "The poison is allowed to fester because there is no antidote. . . . Nobody has the guts to do an intellectually defensible attack [on Bourgois]." The board members were thoroughly annoyed, but the article did not affect their deliberations. They had spent the preceding two days concerned less with the allegations of murder and human rights atrocities by SOA graduates than with adopting new tactics to safeguard the institution that trained them. If they had their way, the School would continue to prepare fresh cadres of soldiers to carry out the custodial duties of empire assigned to them by the United States, and, as the board settled into the task of writing a final report, this was the central thrust of their recommendations.

Sustaining the Momentum

The movement to close the SOA was clearly generating a lot of angst in official circles, but as activists contended with the military's efforts to create a new look for the SOA and to divide and marginalize the movement, they faced a number of tactical and strategic dilemmas. Should the movement concentrate narrowly on the School of the Americas? Or should it focus more broadly on U.S. military interventionism and the U.S. exploitation and domination of Latin American societies? How could it adhere to its religious roots and still reach out to secular constituencies? By grappling with these issues, activists continually rethought their strategies and tactics and, in the process, renewed the movement's identity.[17]

Closing the SOA would be an important symbolic victory, but one that would likely have very little impact on the Pentagon's broader Military Education and Training Program (IMET) and the foreign trainees it brings to numerous military schools across the United States. Abolishing the SOA would also not affect the doctrine that guides the U.S. military around the world. Some in the movement initially saw the School as a blight on the U.S. Army that was not representative of the institution as a whole, and this view was articulated by certain high-profile critics. Actor Martin Sheen, star of the television drama *The West Wing*, lent his celebrity to the movement for several years by heading the procession of line crossers during the November vigil, but the TV commander-in-chief angered some movement activists in 1999 by proclaiming, in the name of those present, that the U.S. Army was "a noble institution" with a "proud tradition" and that the movement's only complaint was with the School of the Americas.[18] Another critic and movement gadfly, retired army major and former SOA instructor Joseph Blair, spoke out repeatedly against the SOA in public addresses and in a movement-produced video, but in these pronouncements, Blair, too, did not link the SOA to broader processes of U.S. military aggression.[19]

Yet many other activists, particularly veterans of previous social justice struggles, saw a more serious issue that transcended the School of the Americas. The real problem, in their view, was U.S. political and economic hegemony in Latin America and the post–cold war world. For these people, what needed changing was not just the SOA but the fundamental nature of U.S.–Latin American relations and the way that the United States operated on the world stage. As the movement developed over the

1990s, a gradual awareness grew among many people that the SOA was, indeed, just the tip of a vast iceberg, and this realization posed a serious challenge to the movement.

On the one hand, if the movement framed the problem as U.S. military interventionism and hegemony in Latin America, how, in a practical way, could it address such an enormous issue? Even though closing the School of the Americas was a major task, it was one that activists could easily imagine, and the institution provided a tantalizing target with potent symbolism. Doing away with U.S. imperialism was another matter. It was an unwieldy undertaking and represented a political project that was simply too big and amorphous to handle in the context of late-twentieth-century America. Analyses of U.S. empire building, military intervention-ism, and regional hegemony were typically marginalized by the main-stream corporate media and viewed as too radical by a broad cross-section of the U.S. public. Defining Latin American dirty war atrocities com-mitted by U.S.-sponsored security forces as the logical outcome of the behavior of a predatory, imperial power and not the result of the flawed policies of a particular U.S. president, or the actions of "rogue" actors, clashed with the beliefs of many American citizens. Posing the problem in this way, especially amid post–cold war U.S. triumphalism and the ascen-dance of free-market orthodoxy, risked placing the movement well out-side the narrow parameters of Democrat-versus-Republican debate and isolating it from broader support.

On the other hand, if the movement focused narrowly on the SOA and the murders of church people and innocent civilians perpetrated by some of its graduates, activists might succeed in closing the institution but fail to have any impact on U.S.–Latin American policy and the way that the United States trains foreign militaries. To make matters worse, even though the SOA continued to teach Latin American soldiers how to kill, despite the absence of institutional accountability for human rights viola-tions, the most controversial aspects of the SOA curriculum had probably moved to other U.S. military schools and abroad, where they were beyond the oversight of the movement and out of the public eye. An exclusive focus on the SOA therefore threatened to make the movement irrelevant, as other parts of the U.S. military apparatus took up the slack from the SOA.[20]

Activists debated the strengths and weaknesses of different strategic approaches, and beneath their shared opposition to the School of the

Americas, different views about tactics and strategy coexisted. Nobody advocated renouncing the immediate objective of closing the School, which was, after all, the sine qua non of the movement, and SOA Watch continued to identify the institution as a "school of assassins," ridiculing the name-change ploy with the slogan "new name, same shame." Yet SOA Watch also began to tie the School and its graduates to other issues, particularly the repressive practices that accompanied economic globalization and that sustained the harsh dictates of international financial institutions, and it laid out these connections in a video titled "Guns and Greed." Drawing out these links enabled SOA Watch to educate movement sympathizers about the School's place in a much broader political and economic context and to grow by building alliances with the burgeoning global justice movement.

Looking beyond the SOA was prompted to a considerable degree by the global justice movement, which burst into public consciousness with the so-called Battle of Seattle in November 1999, when thousands of protesters disrupted a meeting of the World Trade Organization. Four months later, when the protests shifted to Washington, D.C., students organized by SOA Watch participated in the huge April 16, 2000, demonstration against the free-market policies of the World Bank and International Monetary Fund. Through an alliance with the SOA Watch old guard, the students inserted the concerns of the anti-SOA campaign into the broader April 16 action, and they opened more space for themselves in the movement to close the School of the Americas. It was a crucial moment, when, according to activist Bruce Triggs, people in SOA Watch started to "look beyond that one training site. Human rights are threatened by other programs, like the Green Berets training abusive Colombian and Mexican soldiers. We're analyzing the larger system of poverty that soldiers in Latin America defend using military aid and training" (2002, 51).

Youthful protesters brought more than the concerns of the global justice movement with them to the anti-SOA campaign. According to one activist, many of the newcomers also wanted a "less top-down way of participating in the [anti-SOA] protest" and pushed for the formation of base-level "affinity groups." The idea of an affinity group was that, in the context of a large demonstration, small clusters of people take responsibility for each other by making consensual decisions and looking out for companions if confronted with police harrassment or arrest. Many of these young people had already formed affinity groups to participate in earlier protests in Seattle and Washington, D.C. Aware of how the groups mobilized at these venues and demonstrating their capacity to learn from the youngsters, SOA Watch encouraged demonstrators to come to Georgia in affinity groups for the 2000 demonstration.

As more young people joined the struggle, new tensions emerged. Not all the newcomers identified with the movement's religious orientation, and some found it cloying, constraining, and exclusivist. Many in the old guard, however, wanted to ensure that the movement adhered to its roots. These sentiments emerged over disagreements about particular protest styles at the November vigil. In 2000, the bevy of giant puppets that arrived with young demonstrators and included an Al Gore–George Bush hydra, a giant yellow head with "Revolution" emblazoned across the forehead, and a malevolent Uncle Sam, commanded as much media attention as the funeral procession, which had long been the centerpiece of the protest. Moreover, SOA Watch's insistence that marchers treat the funeral procession with reverence and solemnity clashed with the louder, more exuberant style of others who were grounded in anarchist organizing principles and found any hint of rules and hierarchy too restrictive.

At least one elderly activist was unsettled by what he described as "grotesque puppets" and the "sarcastic commentary" of the young. "A lot of these kids," he observed, "are coming from where they are, and we [the old guard] have to tolerate that." Although he acknowledged that everyone was committed to nonviolence, he felt that many young people lacked the spiritual training and depth of the older activists. "Boy," he exclaimed, "they can sure get verbally violent. If you are going to change people's hearts, this is not the way to do it." He then joked about an event that left him and Roy Bourgois befuddled. "A young woman," he recalled, "bared her breast to a police officer, and then said 'fuck you' to someone who told her that the behavior was inappropriate." He chuckled, remembering a phone call that he received from Bourgois. " 'How are we going to handle this problem?,' Roy asked. I told him, 'Roy, you don't handle anything.' "

Young people who chose to trespass onto Fort Benning dealt with the consequences in ways that many of the older, religious activists did not confront so directly. Leone Reinbold, for example, faced prosecution on federal trespassing charges after she and her eleven-member affinity group marched onto Fort Benning with an indictment that they intended to deliver to the School of the Americas. As she awaited trial several months after the protest, Reinbold said that she had felt a strong desire to carry on the movement legacy of crossing the line. "I don't like the idea of the military scaring us out of our history and tradition," she explained. Yet faced with the possibility of a six-month prison term, she admitted that she was having second thoughts about her decision. "If it had happened last year, it would have been fine. I would have said, yeah prison, no problem," she insisted, but the possibility of jail time now seemed like "bad timing." Reinbold was young and single. She rented a nice house in Oakland and was about to be promoted to a new job with higher pay. If she went to jail, she would lose her promotion, her job, and probably her house, and when we spoke, she was entertaining the humiliating possibility of apologizing to the judge in order to get a suspended sentence. The consequences of a prison term for many of the older, religious activists were far less onerous. Some of these individuals were retired people, who received pensions and had spouses to manage their domestic affairs while they were away in jail. Others had the support of their religious orders to which they returned after prison to continue their work.

As young and old, secular and religious, and pacifist and anarchist tendencies struggle to accommodate each other, assessing the movement is a difficult task, as the struggle to close the SOA continues to develop and change. To date, the movement has not managed to close the School of the Americas, but it has scored some notable successes. As the preceding discussion demonstrates, the movement has made military training at the SOA much more difficult by linking the School's training practices to the behavior of its graduates in Latin America and mobilizing public outrage against them. The movement has also forced policymakers to enact a series of reforms to safeguard the institution, reforms that preserve its character as a military training establishment but set it apart from other military schools in the United States that have avoided public controversy. In addition, SOA Watch and the legions of protesters who speak out against military training have waged an important fight against the widespread impunity that characterizes the militaries of the Americas and the historical amnesia they perpetuate. By keeping the names of SOA victims alive and focusing outrage on perpetrators and the institution that trained them, they make it harder to forget the brutality that underlies the maintenance of U.S. interests in the Americas.

Finally, the anti-SOA movement has updated a long tradition of antiwar activism in the United States. It brings new activists into the struggle against U.S. militarism and interventionism in Latin America, even as it bridges the divide with the Central America solidarity movement of the 1980s. These young protesters rub shoulders and learn from an earlier generation of war resisters and social justice activists who continue to fight for peace and equality. The collective insights and solidarity of these diverse groups will be important as the movement enters a new, more complex, and potentially more dangerous phase in post–September 11 America.

The School of America

Great empires die from indigestion.
—Napoleon Bonaparte

Congressional amendments to shut down the School of the Americas (SOA) have consistently failed, once by as few as ten votes, but members of Congress did support a name change in an effort to distance the institution from its cold war legacy. Yet it should come as no surprise that when, in 2001, the School was renamed the Western Hemisphere Institute of Security Cooperation (WHINSEC) very little actually changed. The shift from SOA to WHINSEC represented less the birth of a new training institute than the partial victory of a social movement that brought intense pressure on the military to close an institution implicated in some of the worst human rights violations in Latin America. The new name and the revised curriculum distinguished the Western Hemisphere Institute for Security Cooperation from its more virulent cold war manifestation. Yet the reforms, which took place under pressure over several years, hardly suggested a new mission for the school, and no one was held accountable for the methods taught at the SOA or the behavior of many of its graduates. The SOA-WHINSEC remained dedicated, as it always had been, to training new cohorts of officers ready to defend the ramparts of the American empire.

I have characterized the United States as an empire throughout this book less for the sake of polemics than to capture the way that U.S. power is enacted in the Western Hemisphere. Against some contemporary notions that view empire as a decentered, dispersed apparatus of rule with no territorial core of power (Hardt and Negri 2000), I have discussed a particular empire through the lens of the School of the Americas and given it a home address in the United States. Empire, in this sense, differs from the notion of "superpower," which, as Panitch observes, simply suggests that the United States has more power than other states. The United States is not only the most powerful state in the world. It also uses its power to penetrate and transform other states for its own purposes (Panitch 2002, 16), and this process, I argue, is a key feature of U.S. imperialism. The gradual weaving together of dispersed national armed forces at military centers like the SOA facilitated the large-scale intrusion of U.S. power into Latin America, and it internationalized the repressive capacity of Latin American states.

The military represented the most basic form of imperial control in the Americas. Maintaining U.S. hegemony depended on soldiers to uphold a particular form of capitalist order that shifted over time, and it was a fundamentally violent process. Because of the strategic importance of Latin America for the United States, the School of the Americas played a vital role in training over sixty thousand troops between its founding in the Panama Canal Zone after World War II and its reincarnation as the WHINSEC in 2001. The School constituted part of a hydra-headed, repressive apparatus that included armies, police forces, paramilitaries, training centers, arms manufacturers, and think tanks. This ravenous beast consumed ever more public resources as the cold war spurred its expansion throughout the middle decades of the twentieth century. Massive militarization intensified the demand for professional military officers to oversee permanently mobilized forces, and the predominance of U.S.-supported military dictatorships from the 1960s through the 1980s encouraged some young men from the middle classes to seek social mobility and political power through careers in the officer corps. Even with the termination of the cold war, a military career—complete with training at a U.S. military establishment—enabled some officers to retain a grip on the privileges associated with the middle class at a time when the economic instability that accompanied the advent of unrestrained, free-market capitalism jeopardized the middle class's very existence. The military remained important as neoliberal policies, promoted by the International Monetary Fund and enacted by civilian governments, generated widespread discontentment and rising levels of social decomposition from Mexico to Argentina.

As "governability" became a buzzword in policy-making circles, civilian state administrators continued to rely on the armed forces to control the disorder that they themselves frequently created.

Even though threats, political and economic manipulation, and brute force were key to sustaining U.S. power in the Americas, the new opportunities offered to rising members of the Latin American officer corps by the United States secured their collusion in the U.S. imperial project to a considerable degree. The SOA represented much more than a combat training center for the gendarmes of the domestic status quo. It incorporated upwardly mobile officers into a transnational world of power and privilege, and it presented them with possibilities for advancing their social and economic interests at home. The process of building an imperial military unfolded in a number of ways.

The SOA exposed Latin American trainees to the expert knowledge of its military specialists and the awesome technological capacity of the United States. Technology constituted an important gauge to evaluate the worth of societies and the value of people within them, and the army flouted its technological sophistication as evidence of the innate superiority of the United States and its citizens. Many Latin American trainees believed that access to this quasi-magical world of technology and esoteric information could fortify their own position vis-à-vis local competitors for power, and, indeed, it did. Yet it also bound them closer to the United States, opened them to greater manipulation by the northern behemoth, and preempted military assistance from other states that might challenge U.S. dominance.

The School of the Americas further tantalized students with a particular version of the "American way of life," defined by the commodity-filled, suburban lives of the white middle class, and the School provided some of them with access to it. CGS officers and their families who spent a year or more at the SOA enjoyed experiences that were unavailable to many of their class peers in Latin America. They participated in a comfortable, consumer-oriented world in the Panama Canal Zone, and later, the United States. The opportunity to learn English, to educate their children in U.S. schools, to earn at least some of their salaries in crisis-proof U.S. dollars, and to acquire cheap commodities for personal consumption or sale as contraband enabled them to distinguish themselves from many of their countrymen. Foreign travel, especially to the United States, and the exotic experiences that accompanied it served as an important symbol of a modern, cosmopolitan identity. In addition, completing a course at a prestigious U.S. military school confirmed the students' worth to their armed forces, facilitated their march through the ranks, and gave them an advantage in internal power struggles over those who trained elsewhere. When these perquisites were combined with the sense of professional entitlement instilled at the SOA, as well as in national military academies modeled on U.S. service schools, it should come as no surprise that many in the officer corps viewed themselves as separate from, and frequently superior to, civilians. In some countries, the emergence of separate neighborhoods and social clubs for officers and their families reinforced this detachment. These exclusive settings aggravated the polarization of the world into "our people" and "the enemy," and they intensified forms of racism and class exclusion that were already widespread.

Perhaps most importantly, however, members of the officer corps distinguished themselves from other nouveau riche class segments by the monopoly that the armed forces claimed over the legitimate use of lethal violence. They used arms and combat training to protect the interests of the United States and its local allies in the dominant classes and to destroy the social worlds of those who challenged the basis of class privilege. Domestic security forces consistently disrupted the organizations that peasants, workers, religious, and other ordinary people created to press claims on more powerful groups. They murdered and tortured dissidents and suspected guerrilla collaborators; defended unfair labor practices and inequitable land tenure arrangements; and left a legacy of misery and simmering anger that continues to disrupt the consolidation of stable democracies in several countries. In this way, the rise and consolidation of an armed, castelike group of local allies facilitated the internationalization of U.S. power in the Americas, and it intensified violence and processes of social differentiation locally.

The growth of a hemispheric military apparatus depended on the asymmetrical, personal relationships established between U.S. and Latin American officers at military training centers like the SOA. These men forged personal ties in the Command and General Staff Officers course, where U.S. military officials accorded considerable importance to creating close, working relationships among the leaders of the armed forces of the Americas. The enormous attention paid to developing transnational, interpersonal networks even moved one colonel to complain that "the SOA's substantive curriculum has . . . been only a vehicle for the larger political-diplomatic mission" (Demarest 1994, 49). Through the CGS course, U.S. trainees destined for Latin American postings made contact with the future heads of Latin American militaries early in their careers and spent a year training and socializing with them. Race, nationalism, and rank infused the emergent relationships with tensions that were never completely resolved. Yet a sense of common purpose forged in the battle against variously labeled "internal enemies" bound graduates of particular cohorts, as well as alumni in general, and it facilitated the entrée of U.S. personnel into local armed forces. This access served to promote U.S. policies, and it enabled highly mobile U.S. officials to exert greater control over local security forces. All of this encouraged the gradual transformation of the Latin American militaries into accessories of U.S. power and blurred the boundaries that separated the U.S. military from its regional allies. Referring to Central America during the cold war, Holden maintained that "it is best to speak of U.S.–Central American military power rather than U.S. military power alone, as if it were still exercised independently" (1993, 284).

The expansion of state violence in the Americas rested on the privileges of impunity for both Latin American security forces and their brethren from the United States. Impunity reinforced U.S. hegemony in a number of ways. It enabled the U.S. and Latin American militaries to get away with murder, torture, and the disappearance of thousands of people during the numerous dirty wars that wracked the continent during the last half of the twentieth century. It then shielded them from subsequent demands to clarify and account for their behavior. Impunity also empowered civilian elites, government bureaucrats, and international financial organizations to enact destructive social and economic policies that provoked widespread misery. The impunity-backed state terror that fractured countries such as Guatemala, El Salvador, Chile, and Argentina while they were ruled by harsh regimes set the stage for the subsequent consolidation of neoliberalism under civilian governments (Hollander 1997; Vilas 1996). After the return of civilian rule in the 1980s and 1990s, repressive entities often remained intact, and, because perpetrators suffered negative consequences only occasionally, survivors continued to live in fear (Feitlowitz 1998; Hollander 1997; Green 1999). Speaking out against what had happened, and what continued to happen, remained difficult, and the violent demolition of grassroots organizations and local-level social relationships impeded the ability of ordinary people to take care of themselves and to press their demands on the state and powerful international financial institutions. By the 1990s, social tensions and indices of inequality in Central America were as bad or worse than those reported prior to the outbreak of conflicts in the 1970s and 1980s (Vilas 1996). Crime increased dramatically. As domestic economies were pried open by the International Monetary Fund, a dizzying array of commodities became available to Latin Americans at the same time that the neoliberal economy undermined the ability of many people to purchase them. Consumerism became an invitation to steal, according to Eduardo Galeano (2000), and common criminals came to represent the new "internal enemies" of the security forces, as concerns about public security intensified. Not surprisingly, vigilantism became a problem in areas where the state would not or could not protect people, and, as aggrieved people sought their own justice, the lines between political violence and common crime were more difficult to discern.[1]

Because impunity is the outgrowth of vastly unequal power struggles, its beneficiaries must constantly reinforce, justify, and defend it against challenges from various quarters. The United States continues to protect military perpetrators[2] and the SOA's human rights program must be understood as one part of the struggle to uphold impunity. It is less about human rights than re-imposing the military's definition of reality on the past, present, and future of the Americas. The program skirts serious analysis of the Latin American dirty wars, and it obfuscates the broader imperial history of the United States by presenting a case study of the My Lai massacre that failed to critically address issues of justice for the Vietnamese victims and accountability for U.S. servicemen who ordered, carried out, and covered up the slaughter. By appearing to engage the issue of human rights, the program attempts to silence critics of the SOA, while it obscures the violent twentieth-century history of the Americas and justifies the continued training of unreformed Latin American militaries at a time when the School's institutional survival is under threat.

Most of the upwardly mobile students of the SOA's prestigious Command and General Staff Officer course are overtly contemptuous of human rights organizations, which they accuse of besmirching the image of the Latin American armed forces, and they do not hesitate to express their views when confronted by these groups. Although some took the SOA's human rights program more seriously than others, the forty-hour series of lectures and seminars cannot possibly alter beliefs and practices that are deeply ingrained in military institutions. Basic respect for a broad range of fundamental entitlements and freedoms transcends the classrooms of the SOA. How rights are conceptualized depends on the shifting balance of social, economic, and political power in particular societies, especially within their armed forces, at specific moments in time. Because Latin American militaries remain unreformed and generally unaccountable for the death, terror, and destruction inflicted on unarmed civilians in the past, it should come as no surprise that their trainees at the School of the Americas are scornful of human rights organizations and express no remorse for the behavior of their institutions. The same is true for many members of the U.S. military who teach them. Furthermore, in the Andean countries of Colombia and Bolivia, where states are fighting their own people, earning the privilege to attend the School of the Americas depends on the numbers of arrests, eradicated coca fields, destroyed cocaine pits, and, in Colombia, dead bodies produced by soldiers and militarized policemen in their battles to halt the drug traffic and suppress guerrilla insurgencies. These production pressures only aggravate human rights violations in countries where impunity for the armed forces is firmly entrenched.

Even though impunity has shielded killers and upheld a particular form of capitalist political and economic order in the Americas, the order that it sustains is extremely unstable and requires arms, military training schools, a disregard for human rights, and constant vigilance to maintain. Order and disorder are thus intimately related. This is most evident in the coca-growing regions of Colombia and Bolivia, which are the contemporary focal points of U.S. interventionism in the Americas. Impunity-generated political and economic violence ruptures the lives of ordinary people, leaving fear, increased poverty, destroyed social relations, and death, and alternative development programs have not incorporated poor settlers into new productive arrangements. Alternative development is merely the benign face of a repressive drug war that seeks to deprive peasants of their only viable cash crop. Today, what is at issue for these people, as Sider and Smith suggest for another context, "is not the powerful interested in order and the weak opposed to it, but . . . the weak seeking (their) order against the chaos imposed by domination" (1997, 24). Because state bureaucrats are unwilling to address the basic needs of coca-growing families, peasants have little choice but to continue growing coca and, in the case of Colombia, forging alliances with illegal armed actors for protection. All of this, in turn, prompts additional violence by states to control conditions that are to a considerable extent their own creation and beyond their capacity to manage completely. This violent and contradictory quest to impose order arises from impunity, and is molded by it.

The empire is much more dangerous and unstable today than it was prior to the September 11 terrorist attacks, and the word "empire" is no longer stifled by the decorum that long precluded its utterance by mainstream pundits on the op-ed pages of major newspapers and magazines. The E-word is now everywhere, and discussing empire and imperialism is respectable enough. Writing in *Harvard Magazine*, for example, Stephen Peter Rosen does not beat around the bush. He opines that "the U.S. has no rival. We are militarily dominant around the world. . . . A political unit that has overwhelming superiority of military power, and uses that power to influence the behavior of other states is called an empire. . . . our goal is not combating a rival but maintaining our imperial position, and maintaining imperial order."[3] Similarly, Michael Ignatieff asserts in a *New York Times Magazine* cover story that "there are many people who owe their freedom to the exercise of American military power."[4] The contemporary advocates of empire view the aggressive enforcement of global order by and for the United States as a positive virtue. For Niall Ferguson, the only question is "whether or not America has the one crucial character trait that without which the whole imperial project is doomed: stamina."[5]

The assaults on the World Trade Towers and the Pentagon have unleashed the self-righteous fury of the United States and bolstered those sectors of the state apparatus charged with teaching, controlling, and dispensing violence. The current "war on terror" has no foreseeable end, and it has provided new justification for remapping the world in accord with the political, economic, and military interests of the United States. The Middle East, in particular, has become more volatile than ever. The war in Afghanistan dispersed the al-Quaeda terrorist network, and the Bush administration capitalized on domestic fears of terrorism to justify attacking Iraq, laying claim to its vast oil fields, and intensifying U.S. domination in the region.

The politics of oil and terror also shape deepening U.S. military involvement in Colombia, which has become attractive as an alternative source of oil for the voracious American economy. The U.S. government now labels the FARC guerrillas "terrorists," thus replacing older designations—"communist" and "narcoguerrilla"—that lost their timeliness and political expediency after the cold war and September 11 respectively. The U.S. armed forces are moving under the guise of a war on terrorism to intervene more directly in the Colombian conflict. In early 2003, the deployment of U.S. Special Forces trainers to the Colombian department of Arauca, where they provide counterinsurgency instruction to local troops, forms part of a $94 million counterterrorism aid package designed to protect the 478-mile-long Caño Limón oil pipeline used by Los Angeles–based Occidental Petroleum.

Closer to home, the U.S. Army uses the tragedy of September 11 to reaffirm the soa-whinsec's mission. According to the army, it is more important than ever to "engage" with Latin American militaries, as the threat of terrorism rises. Additional monies for international military education now permit more students to attend the School than prior to September 11, 2001, and army officials expect the annual enrollment to reach one thousand. Talk of terrorism is on everyone's lips, and Latin Americans may soon be offered a course on the topic.[6] Yet despite the real threat of terrorist violence from the shadowy al-Qaeda network, the "terrorism" label has already become a catchall marker that ostensibly identifies and delegitimizes a variety of dissenters.

Even more disturbing, September 11 has altered the moral radar of broad sectors of the American public. The legitimacy of torture is now discussed on prime-time network television,[7] where conversations about its utility were unimaginable prior to the terrorist attacks. Moreover, torture is being redefined to exclude practices that it previously included, such as sleep deprivation, psychological manipulation, and forcing victims to maintain uncomfortable positions.[8] Torture brutalizes the tortured and degrades societies that approve of it. Yet army interrogators now acknowledge publicly the use of torture on suspected terrorists. An official who oversaw the capture of alleged Middle Eastern terrorists told *Washington Post* reporters Dana Priest and Barton Gellman, "If you don't violate someone's human rights some of the time, you probably aren't doing your job. I don't think that we want to be promoting a view of zero tolerance on this."[9] In addition, American military officials admitted in March 2003 that two prisoners captured in Afghanistan died from "blunt force" injuries while under interrogation at the Bagram air base. The official cause of death was homicide.[10] To what extent, we might ask, would a course on terrorism at the soa-whinsec simply update the brutal methods used to fight the old communist boogeyman? And what would prevent alumni from using their training against the same kinds of men, women, and children who bore the brunt of the twentieth century's dirty wars?

The answers to these questions will not come from U.S. Army officials. When I spoke with the new commandant, Colonel Richard Downie, about the impact of September 11 on the mission of the new institution, he quickly informed me that he could not speak about the School of the Americas, even though I had not mentioned its name. Yet Downie commanded the SOA-WHINSEC in large part because of his extensive experience in Latin America, especially Colombia, Panama, and Mexico, and it was inconceivable that he could not talk about the SOA.[11] Denying any knowledge of the SOA was simply the latest ploy used by the military to separate itself and the WHINSEC from the crimes of the cold war and the methods taught at the School. There was nothing particularly new about this ruse. Disappearing history was a well-developed strategy used by U.S. and Latin American military officials to avoid any responsibility for past human rights violations and to recast the legacy of U.S.-Latin American relations in the present. For besieged officials at Fort Benning and in the Pentagon, it was also part of an ongoing effort to diffuse a social movement that demanded an end to state-sponsored terrorism by the United States.

The September 11 attacks handed the army and its civilian supporters in the local government a new weapon in the campaign against Father Bourgois and the thousands of demonstrators who converge on Fort Benning every year to commemorate the victims of SOA graduates and to call for the closure of the School. The military and its local civilian boosters had long sought a way to portray nonviolent protesters as threats to public safety. In 2000, for example, the army circulated rumors about "anarchists from Seattle" who planned to come to Georgia and sow havoc during the protest vigil. In 2001, two months after the destruction of the World Trade Center and the attack on the Pentagon, it capitalized on domestic fears of terrorism. The city government filed an injunction against protest organizers, threatening them with prison if they proceeded with the demonstration. Although an eleventh-hour court order overturned the injunction, the terrorist fearmongering continued the next year, when the demonstration was even larger. City officials forced over ten thousand activists to pass through metal detectors before entering the vigil site, while a handful of pro–Fort Benning counterdemonstrators were spared the indignity, because, as an official told me, "they are standing on private property." Such aggressive efforts to intimidate anti-SOA activists highlighted the growing intrusiveness of the state security apparatus in the daily lives of U.S. citizens and the increasing fragility of civil liberties in post–September 11 America.[12]

The military no longer underestimates the tenacity of the anti-SOA resistance. Although it effectively divided the congressional opposition by changing the name of the institution, the thousands of ordinary U.S. citizens who form the rank and file of the anti-SOA movement have not been dissuaded so easily, and they continue to call for the closure of what they call a "School of Assassins." Although to date the SOA-WHINSEC continues to operate as a military training center, and the U.S. military is unrepentant about its eagerness to work with militaries that persist in committing human rights violations, the movement has forced the School to adopt a series of reforms not seen at other U.S. military training centers that have avoided public controversy. Manuals used at other institutions are now sanitized before they are adopted at the School, and course offerings have been sterilized and restructured in an effort to portray the SOA-WHINSEC as a more civilian-friendly institution.

Anti-SOA activists have also undermined the military's efforts to rewrite the history of the Americas. Their yearly vigil at the gates to Fort Benning keeps alive the memory of the dead and the disappeared, and by so doing, it chips away at the structure of impunity that the armed forces of the Americas struggle so mightily to uphold. Moreover, as rising nationalist sentiments within the United States fuel an open-ended war on terrorism, the movement provides an alternative perspective and encourages people to imagine a different kind of nonviolent world in which justice is available to everyone.

Yet the movement now finds itself at a crossroads, where it is threatened with becoming a victim of its own success. Although the SOA-WHINSEC is an important symbol of atrocities perpetrated by the security forces of Latin America, it is neither the only nor the most important center where military training takes place, and because of the intense public scrutiny brought by legions of activists, the most sinister aspects of training are gone, relocated in all likelihood to centers south of the border and less controversial, domestic military schools. Although the movement may succeed one day in closing the SOA-WHINSEC, the policies that gave birth to the SOA at the dawn of the cold war, and then spawned the invasions of Cuba, the Dominican Republic, Grenada, and Panama, generated the Contras, trained death squads, and supported right-wing military dictators from Ríos Montt in Guatemala to Pinochet in Chile, will remain in place. And today, these policies are more virulent than ever. Without ending U.S. imperialism and dismantling the military apparatus that supports it, many in the movement understand that their government will almost certainly continue to ride roughshod over the rights of Latin Americans. Anti-imperialist arguments, however, are only made in certain quarters of the movement, and they are usually not captured in official slogans, such as "Not in Our Name" and "Close the School of the Assassins." Some movement activists see a positive part for the U.S. military to play in Latin America and object only to what they view as the aberrant behavior of the SOA, particularly the murder of priests and nuns by its graduates. As one middle-aged woman from the Midwest told me during the 2002 protest, "I just can't take the assassination training."

Anti-SOA activists of various persuasions will negotiate and renegotiate the tactics and strategies of the movement, as they have done for the last decade within an ever changing political context. This is the substance of grassroots political activism. There is neither a simple nor an obvious political program that can be prescribed for action. Yet as activists strategize about the future, they will likely confront government officials who are less inclined to dress up their policies in Sunday school clothes for public consumption. Indeed, American imperial ambitions find considerable backing among many U.S. citizens, whose deep-seated nationalist and xenophobic sensibilities are aggravated by fears of future terrorist attacks.

All the issues, however, will not be decided in the United States. The story of the School of the Americas and the broader imperial adventure of which it is a part continues to unfold across Latin America, where the chains of empire weigh heavily on many people. Amazingly, some of these men and women always find ways to organize and resist. Their struggles constantly undermine the imperial order that the military trains so diligently to preserve, but as the global reach of the U.S. military apparatus expands, solidarity within and among the social movements of the Americas becomes increasingly important to restoring memories of the cold war past, securing justice and accountability in the present, and building a more peaceful future.

NOTES

✦

Prologue

1 The allegations against Urbina are presented in *El terrorismo de estado en Colombia*. Brussels: Ediciones NCO, 1992.

2 See OMCT 1992 for reference to these allegations.

3 The DAS is an investigative police force whose members operate without uniforms. Although police units are administered by the Interior Ministry, the DAS is run by the executive branch, and under Urbina's command, it was an example of how the police and army intersect.

4 See the collection of declassified documents published by the National Security Archives under the title "War in Colombia: Guerrillas, Drugs and Human Rights in U.S.-Colombia Policy, 1988–2002." National Security Archive Electronic Briefing Book No. 69. Washington, D.C., 2002.

5 Urbina is referring to *El terrorismo de estado en Colombia*.

Introduction

1 See Dinges (2004) for more on the CIA and Operation Condor.

2 See Owen and Sutcliffe (1972) for more discussion of imperialism and the various ways in which the concept has been understood by scholars and activists.

3 See Schirmer (1998) for a discussion of the hold that the Guatemalan military maintains over the political process in that country.

4 For more discussion of this conceptualization of U.S. imperialism, see Panitch and Gindin (2003). See also N. Smith (2003) for an interesting geographical analysis of the emergence of the U.S. empire in the first half of the twentieth century. Smith describes the emergence of a new geographical sensibility among U.S. policymakers that relied less on the old European predilection for

marking colonial boundaries and controlling absolute space than on organizing the social relationships necessary for capitalist development through the abstract notion of "the market"—a form of capitalism that was allegedly "beyond geography," even though it remained deeply rooted in local and national forms of territorial power. The resulting differences in European and U.S. styles of economic expansion fed domestic beliefs about "American exceptionalism" and the American state's insistence that it was not imperialistic.

5 See Panitch and Gindin (2003). Chomsky (2000) provides an interesting discussion of the notion of so-called rogue states.

6 Although U.S. policymakers and many academics have long denied the reality of empire, a number of recent collections seek to restore the study of empire to the social sciences. See Joseph, LeGrand, and Salvatore (1998); Stoler and Cooper (1997); Kaplan and Pease (1996). For a discussion of the analytic utility of the concept "imperialism," see the special issue of *Radical History Review* 57 (fall 1993), "Imperialism: A Useful Category of Analysis?": 1–84.

7 See Holden (1996) for a general discussion of the internationalization of state-sponsored violence in Central America. See Huggins (1998) for an interesting discussion of how the U.S. training of Brazilian policemen expanded the United States' control over the Brazilian security system.

8 This figure comes from Lumpee (2002), but Amnesty International places the number of training centers in the United States at 275 (Amnesty International 2002, iv).

9 In some cases, however, perpetrators did pay a high price for their behavior. Lieutenant Colonel Domingo Monterrosa, who commanded the El Mozote massacre, was killed in a clever ruse carried out by the *Ejército Revolucionario del Pueblo* (ERP), a branch of the FMLN. The ERP led Monterrosa's troops to believe that they had captured *Radio Venceremos*, the rebel radio that broadcast clandestinely throughout the war. Monterrosa flew to the site where soldiers had discovered an old Viking radio transmitter. Unbeknownst to the military, the ERP had rigged the device with a bomb. Monterrosa placed it in his helicopter and then set out for San Salvador, where the press was waiting to report on his "discovery." As the helicopter ascended, a remote-controlled device triggered the bomb, which blew up the helicopter and killed Monterrosa. Monterrosa was an SOA graduate and had also received training from the U.S. military in other locations.

10 This is basically the same strategy adopted by Payne (2000). I had originally considered a methodology more akin to the one utilized by Gusterson (1998) in his study of nuclear scientists at the Lawrence Livermore nuclear laboratory. Gusterson moved to the town of Livermore and then, through his participation in a variety of local social groups, "began at the bottom, arduously assembling my own network of personal contacts within the laboratory until, eventually, I worked my way up to some of the senior managers"

(36). Such an approach, as Gusterson correctly indicates, makes one's research less subject to manipulation by dominant groups. Yet given the frequency with which SOA students pass through Georgia, and the ways in which their own relationships to U.S. society are shaped by the School, I chose not to adopt this approach, despite its considerable strengths. I was also interested in the ongoing connections between the U.S. Army and SOA alumni in Latin America, which obliged me to make several trips to the region over the course of my fieldwork.

11 My experiences interviewing SOA alumni are very similar to those described by Leigh Payne, who interviewed participants of right-wing, "uncivil" movements in Latin America. Her interviewees included former Nicaraguan "contras," Argentine Carapintadas, and members of the violent Brazilian Rural Democratic Union. See Payne (2000, xxv–xxx).

12 Huggins et al. (2002) and Conroy (2000) make similar arguments in their studies of torturers in various parts of the world. Like Milgram, they conclude that men who torture are not barbarians, even though they commit barbaric acts. Their research examines the wider social and political contexts that permit torture and allow torturers to rationalize, justify, and cover up crimes.

13 See Sluka (2000) for a discussion of the issues that confront anthropologists who study the state and the perpetrators of violence.

Chapter 1. Georgia Not on Their Minds

1 The bemused duck who trumpets the virtues of AFLAC in television commercials seen by millions of people belies the intense right-wing political convictions of the company's founders.

2 Minorities constitute 45 percent of army enlisted personnel and 23 percent of the officer corps. African Americans, who constitute 12.8 percent of the U.S. population, are overrepresented among enlisted personnel (22.4 percent) and underrepresented in the officer corps (8.9 percent) (CDI 2002, 25–26).

3 For example, huge protests erupted in Cochabamba, Bolivia, in 2000 when the Bechtel Corporation acquired control of the public water supply and announced drastic price increases that were well beyond the reach of most residents.

4 Between 1956 and 2001, the Georgia state flag showcased the battle flag of the Confederacy. A new flag was officially approved on January 30, 2001, after Georgia Governor Roy Barnes—a centrist Democrat—proposed a redesign of the flag that minimized the Confederate symbolism. The adoption of a new flag was very controversial, however, and Barnes lost his 2002 bid for reelection to Sonny Perdue because of the flag issue.

5 These attitudes date to the second World War, when Puerto Ricans were drafted into the army in large numbers after being made citizens. Officers

tended to trust Puerto Ricans less than continental troops and believed that they were emotionally unstable and unintelligent (see Lindsay-Poland 2003, 57–59). Although class background shapes the nature of spoken Spanish everywhere, Puerto Rican Spanish is very different from the mainland Latin American variants.

Chapter 2. De-Mining Humanitarianism

1 School of the Americas, personal communication.
2 See the database of SOA graduates from Guatemala available at the office of SOA Watch in Washington, D.C., and in the National Security Archives, Gellman Library, George Washington University, Washington, D.C.
3 These soldiers were a fraction of those Guatemalans who received U.S. military training at the SOA and elsewhere. See the database of SOA graduates in Washington, D.C., and in the National Security Archives, Gellman Library, George Washington University, Washington, D.C., for a record of those who attended the SOA.
4 Report of Investigation. Improper Material in Spanish-Language Intelligence Training Manuals. Department of Defense, March 10, 1992. National Security Archives, Washington, D.C.
5 Since its founding, NACLA has in fact published numerous articles critical of the Soviet Union, Cuba, and Latin American communist parties.
6 With the declassification of the SOA graduate list, it became apparent that the School had trained some of the most notorious perpetrators of human rights violations in Latin America. Further declassification of the names of alumni from other U.S. service schools is still necessary to construct a complete picture.
7 See Huggins et al. (2002) for more discussion of how organizational affiliations, secrecy, insularity, and anonymity shape the behavior and moral universes of Brazilian policemen.

Chapter 3. Foot Soldiers of the U.S. Empire

1 "Brief History of Military Missions," Record Group 338, Box 78, National Archives, Washington, D.C.
2 "Request for Historical Summary of Major Accomplishments of Military Missions," Record Group 338, Box 78, National Archives, Washington, D.C.
3 Ibid.
4 "Advisability of Continuing the Latin American School," U.S. Army Caribbean, Record Group 338, Box 329, National Archives, Washington, D.C.
5 See Goñi (2002) for an interesting discussion of how the Perón government helped Nazi war criminals escape to Argentina and then protected them from prosecution.

6 "Special Staff Meeting on 90-Millimeter Gun Instruction for Argentine Military Personnel," Record Group 338, U.S. Army Caribbean, Box 223, National Archives, Washington, D.C.

7 "Staff Study," Record Group 338, U.S. Army Caribbean, Box 223, National Archives, Washington, D.C.

8 "Special Staff Meeting on 90-Millimeter Gun Instruction for Argentine Military Personnel."

9 Ibid.

10 "Staff Study."

11 See Lindsay-Poland (2003, 32–39) for more discussion of the army's racial beliefs in Panama.

12 Ibid.

13 Ibid.

14 "Information for Argentine Officers, Special Antiaircraft Course," Record Group 338, U.S. Army Caribbean, Box 223, National Archives, Washington, D.C.

15 "Staff Study."

16 "Selection of Students for the Latin American Ground School," Record Group 338, U.S. Army Caribbean, Box 222, File 352, National Archives, Washington, D.C.

17 Ibid.

18 "Leadership," Record Group 338, Caribbean Defense Command, Box 334, File 352, National Archives, Washington, D.C.

19 "Evaluation of Latin American Students," Record Group 338, Caribbean Defense Command, Box 329, National Archives, Washington, D.C.

20 "Leadership."

21 "Selection of Students for the Latin American Ground School."

22 "Student Training," U.S. Military Ground Mission, U.S. Embassy, San Salvador, El Salvador. Record Group 338, U.S. Army Caribbean, Box 222, National Archives, Washington, D.C.

23 "Student Training," Chief, U.S. Military Ground Mission, Peru (1947) Record Group 338, U.S. Army Caribbean, Box 222, National Archive, Washington, D.C.

24 These figures and all subsequent figures about the student population at the SOA come from a database of graduates released under the Freedom of Information Act. It is available at SOA Watch in Washington, D.C., and at the National Security Archives, Gellman Library, George Washington University, Washington, D.C. Information on specific graduates known to be implicated in human rights violations is also available on the SOA Watch Web page at www.soaw.org (accessed 30 November 2002).

25 Anastasio Somoza Debayle assumed the presidency of Nicaragua in 1963, the year that Colonel Himes returned to the United States to assume the post of subcommandant at Fort Hood.

26 Interview with General Fred Woerner (ret.), Washington, D.C., 11 November 2000.

27 See Noriega and Eisner (1997) for Noriega's version of his relationship with the United States and the invasion that overthrew him.

28 Examples of notorious Colombian alumni from the 1970s include: (1) General Harold Bedoya Pizarro, who was a guest instructor in 1978–79. Throughout his career, Bedoya has been repeatedly linked to death squad activity but has never been prosecuted. (2) Major Jorge Flores Suázo, who took a military intelligence course in 1972 and was subsequently implicated in paramilitary activity. (3) Captain Héctor Alirio Forero Quintero, who studied small-unit infantry tactics in 1977 and went to command a patrol that disappeared four people in 1988. On the same day, he detained and tortured two other individuals with the help of fellow soa graduate Carlos Morales del Río. (4) Major Hermann Hackspiel Olano, who also learned small-unit tactics and was implicated in the 1988 massacre of twenty banana workers in Urabá.

Examples of notorious Salvadoran alumni from the 1970s include: (1) Colonel Napoleon Alvarado, who graduated from the military police course in 1974 and, in 1983, covered up the Las Hojas massacre in which sixteen civilians were killed. (2) Colonel José Emilio Chavez Cáceres, who took military intelligence and urban counterinsurgency in 1975 and 1974 respectively. He was in command when soldiers murdered ten civilian prisoners in 1988. (3) Lieutenant Yusshy René Mendoza Vallecillos, who headed the patrol that murdered six Jesuits, their housekeeper, and her daughter. (4) General Juan Orlando Zepeda, who trained in urban counterinsurgency operations in 1975. He is cited by the Human Rights Commission of El Salvador for participation in 210 executions, 64 cases of torture, and 110 illegal detentions. (www.soaw.org/grads/)

29 "IMET Policy Guidance." Cable sent by the Military Liaison Officer in Quito, Ecuador, to the Commander-in-Chief of southcom and the United States Military Groups in Latin America. May 16, 1984. Record Group 338, Box No. 556, Accession No. 338-95-556, File Name: Army Program 1984. Washington National Record Center, Suitland, Md.

30 See Huggins (1998) for an interesting discussion of this program and an analysis of how it operated in Brazil.

31 "Honduran G-2 Enroute to South America," Department of Defense Declassification Document #16, Batch DC38, National Security Archives, Washington, D.C.

32 After the existence of Battalion 3-16 became public and López Grijalva was implicated in its activities, he was still invited to speak at the School of the Americas in 1991 and 1992.

33 The same is true of other countries. See, for example, Richani (2002) on Columbia and Huggins (1998) on Brazil.

Chapter 4. Pathways to Power

1 A variety of declassified documentation about Operation Condor is available on the Web site of the National Security Archive (www.gwu.edu/~nsarchiv/). Accessed 30 November 2002. See also Dinges (2004) and McSherry (1999).

2 This practice was deadly serious. The military demolished the COB headquarters in 1980, after first attacking it and killing or imprisoning those it apprehended. These events occurred in the aftermath of General Luis García Meza's "cocaine coup" and on the orders of Colonel Luis Arce Gómez.

3 Otro conocimiento que me acuerdo muy bien, que trajeron de la Escuela de las Américas, y que era como axiomático de los Rangers, como que mejor tener un subversivo muerto que tenerlo prisionero. Tenerlo prisionero interfería las operaciones siguientes, entonces es mejor que esté cuatro metros bajo tierra que tenerlo vivo.

Chapter 5. Strategic Alliances

1 See the SOA graduate database available at SOA Watch, Washington, D.C., and the National Security Archive, Gellman Library, George Washington University, Washington, D.C.

2 Trumble's background was with the Special Forces, and before coming to the SOA, he worked in the 1970s and 1980s with Special Forces units that operated out of the Panama Canal Zone. Sometime after leaving the SOA, Trumble took another military assignment in Colombia, where he worked in 2001.

3 As early as the 1970s, the Defense Department responded to charges that it taught inappropriate subjects to Latin Americans by modifying the titles of courses that it offered in training schools in the Panama Canal Zone (Schoultz 1981, 232–33).

4 The Kaibles were widely known for extreme human rights violations. (See, for example, Manz 1989.)

5 This name is a pseudonym.

6 This name is a pseudonym.

7 According to Greg Grandin, the Guatemalan state began to promote indigenous culture as a tourist attraction as early as the 1930s. At that time, state and military elites began appropriating it as "folklore" for their own purposes (2000, 194–95).

Chapter 6. Human Wrongs and Rights

1 See SOA Watch's list of notorious graduates at www.soawatch.org.

2 See Carbonella (2001) for an interesting discussion of memory and the figure

of the Vietnam veteran in the construction of American nationalism and in thinking about the United States as an empire.

3 See Huntington (1981, 7–18).

4 For example, during the U.S. invasion of Panama, the fires that burned down the Panama City neighborhood of El Chorillo were probably started by errant rockets fired from an army helicopter at Panamanian military head-quarters. For an interesting critique of the just-war thesis in the aftermath of the September 11 attacks on the United States, see Zinn (2001).

5 Zinn (2001) argues that just-war proponents, such as Richard Falk (2001), confuse a "just cause" with a "just war." While there are, indeed, just causes, such as ending terrorism, "it does not follow," states Zinn, "that going to war on behalf of that cause, with the inevitable mayhem that follows, is just."

6 See Leigh Payne's discussion (2000) of right-wing social movements in Latin America for an interesting analysis of how movement activists used just-war analogies and military symbolism to justify their violent activities.

7 In other contexts, however, political considerations dictate against using the United States as a case study. A segment of the CGS course that focused on geopolitics excluded examples of the United States. The instructor explained that geopolitics refers to the ways that states grow and expand. He preferred not to use the United States as an illustrative case because of the sensibilities of many students, especially the Mexicans who resent the loss of territory to the United States.

8 At least one veteran's group had urged him not to participate. When I later encountered Thompson outside the building, taking deep drags on a ciga-rette during a break, he said to me, "You know what this place is don't you? It's one hundred years of progress without reform."

9 The 1955 Supreme Court ruling in *Toth v. Quarles* states that former members of the armed services are immune from prosecution in military courts "no matter how intimate the connection between the offense and the concerns of military discipline."

10 See Lumpee (2002).

11 In 1997 the U.S. Congress included a special amendment in the Foreign Opera-tions Appropriations Act that was designed to prevent international antinar-cotics funds from going to units in the Colombian military known to violate human rights. The amendment was known as the "Leahy Provision," after its author, Vermont Senator Patrick Leahy. In 1998 a similar restriction was applied to the Defense Appropriations Act that applied to military training and included IMET monies used to pay for instruction at the SOA and other U.S. military schools. This legislation was important, and according to Hu-man Rights Watch, it has had an impact in Colombia. The measure, however, has not ended human rights violations in the Colombian military, because, as Human Rights Watch indicates, "it is often impossible to know the names of

individual security force members who are alleged to have committed violations" (2001, 81).

12 While the coca bush is not destructive of the environment, the social and economic pressures that drive peasants to clear increasingly more tropical forest to plant coca do harm to fragile tropical ecosystems. In addition, the practice, associated with the cocaine industry, of dumping chemicals into rivers and streams also damages the environment.

Chapter 7. Disordering the Andes

1 President Nixon first declared war on drugs in the 1970s.

2 See Melguizo (2001) for an analysis of social disintegration and violence in the Colombian city of Medellín. Also, Goldstein (2002) provides an interesting analysis of vigilante violence in a poor neighborhood of Cochabamba, Bolivia.

3 See Vilas (1996) and Caldeira and Holston (1999) for discussions of democracy and citizenship in Central America and Brazil respectively and rising levels of violence in nominally democratic states.

4 Gerald Sider's initial theoretical discussion of impunity as rooted in the process of social differentiation has informed my thinking about the concept in this chapter. See Sider (2000). See also Priscilla Hayner's assessment of various truth commissions around the world and how they balance the demands for truth and justice (Hayner 2001).

5 See Weil and Weil (1993) for a discussion of how sindicatos organized frontier life in an early frontier settlement in Chapare.

6 See Sanabria (1993); Leóns and Sanabria (1997), and Spedding (1994). See Gootenberg (1999) for an interesting collection of articles that explores the historical roots of the drug cocaine and its transformation from medical miracle to social threat.

7 See Malkin (2001) for a discussion of drug trafficking in rural Mexico and how it was situated within a cyclical history of periodic economic booms and busts.

8 I take this sense of "outlaw" from Peter Linebaugh's interesting account of the social and economic circumstances that shaped particular crimes in eighteenth-century London (1992). The book is a compelling analysis of the relationship between capital punishment and the punishment of capital.

9 In 1999 there were fifteen Special Forces deployments in Bolivia (Isacson and Olson 2001).

10 It was not possible to verify Layme's claims about the Special Forces.

11 In the 2002 presidential elections, Morales scored a huge victory by winning 21 percent of the popular vote. During the campaign, U.S. ambassador Manuel Rocha angered many Bolivians by threatening to cut off U.S. aid to Bolivia

if Morales were elected. His remarks, which many interpreted as meddling, increased support for Morales, according to some observers. Morales's showing placed him at the head of a large opposition block in congress that controlled 26 percent of the 157 seats, a better position from which to challenge the drug war.

12 These testimonies were provided to me by René Laime, the head of Radio Soberana.

13 See International Narcotics Control Strategy Report, Bureau for International and Law Enforcement Affairs, U.S. Department of State, Washington, D.C., 2001.

14 See Leech (2002) for a useful summary of Colombia's economic crisis and the political violence that wracked the country for much of the twentieth century.

15 In my conversations with Colombian military officers, I was consistently struck by their disinterest in the drug war. They preferred to discuss the guerrillas, especially the FARC, and addressed the fight against the drug traffic only insofar as it related to undermining the financial base of the insurgents.

16 Support for a counterinsurgency war was voiced by a 2001 report issued by the influential Rand Corporation which stated that "U.S. policy . . . misses the point that the political and military control that the guerrillas exercise over an ever-larger part of Colombia's territory and population is at the heart of their challenge to the Bogotá government's authority. The United States ought to rethink whether the distinction between counter narcotics and counterinsurgency can be sustained, and whether Colombia and its allies can be successful in the war against drugs if the Colombian government fails to gain control of territory" (Rabassa and Chalk 2001, xviii).

17 See Molano (1992) for more on the emergence of the FARC.

18 The paramilitaries in this sense were very different from their counterparts in Central America, who operated as appendages of the armed forces.

19 In just one year, 1999, the Defensoría del Pueblo documented thirteen massacres in the department, which resulted in the deaths of seventy-seven people.

20 The U.S. government eventually conceded in the spring of 2000 that alternative development in Colombia was a failure and that it would no longer offer assistance to peasants whose coca fields were eradicated. Aerial spraying of coca fields continued, however.

21 This name is a pseudonym.

22 As Chevigny (1995) notes in a study of Latin American urban police forces, a decline in abusive practices is linked closely to the enforcement of viable systems of accountability.

23 See "Human Rights World Report 2001," Human Rights Watch. www.hrw .org / wr2kl / americas / colombia.html. (Accessed 28 October 2002).

24 See the SOA Watch Web page www.soaw.org (Accessed 2 November 2002).

25 "Huanca death 2001," La Paz 04988, Department of State cable, 1 December 2001. Obtained through the Freedom of Information Act. Files of Jeremy Bigwood.

26 Department of Defense cable, October 1997. Document ID: 189695174. Obtained through the Freedom of Information Act. Files of Jeremy Bigwood.

27 According to Human Rights Watch, it is not unusual for the paramilitaries in Putumayo to pay victims' families for their "errors" (HRW 2001, 22).

28 See Meertens (2001) for more discussion on how displacement affects women.

29 See Gill (1997) for more discussion of compulsory military service and the local-level alliances and divisions that it creates.

30 See Alvarenga (1996) on how these processes played out in turn-of-the-century El Salvador; Paul and Demarest (1988) and Godoy (2002) for contemporary Guatemala; and Gourevitch (1998) for Rwanda.

31 Personal communication. Teófilo Vásquez, Bogotá, Colombia. January 19, 2002. See also Molano (1992).

32 Nationwide protests and road blockades in the Chapare in September and October 2000 persisted for weeks and exacted a heavy toll on the Bolivian economy. The government had learned its lesson and was not going to allow a repeat of these events.

Chapter 8. Targeting the "School of Assassins"

1 In the 1980s, Bourgois and a small group of activists staged a protest against the training of Salvadoran soldiers at Fort Benning. Impersonating army officers, they entered the base at night and made their way to the barracks that housed the Salvadorans. They scaled a tree and installed large speakers on the branches. Then, after the lights went out inside, the men blasted the final sermon of Salvadoran Archbishop Oscar Romero from a boombox. Given in the cathedral in San Salvador at the height of the repression in the early 1980s, the sermon admonished soldiers to stop the killing. Romero was murdered shortly thereafter on the orders of SOA graduate and death squad organizer Roberto D'Aubuisson.

2 Smith (1996) explores the importance of religious faith in sustaining Central American solidarity activists and in shaping their understandings.

3 For more discussion and analysis of the sanctuary movement, see Cunningham (1995).

4 See Taylor (2000, A3–A4).

5 Smith (1996) examines the role played by Witness for Peace and other progressive faith-based organizations in the Central American peace movement.

6 In recent years, Witness for Peace and SOA Watch have also cosponsored special delegations to Colombia for anti-SOA activists who have served prison sentences.

7 See Ronfeldt et al. (1998) for a discussion of how the military views the use of the Internet by political activists. They refer to the phenomenon as "social netwar" and present a series of ideas about developing "counternetwar" tactics.

8 See "Terrorism and the Urban Guerrilla," U.S. Army training manual, 112. National Security Archives, Washington, D.C.

9 This tactic was also used against African American civil rights protesters by the police in Albany, Georgia, in 1961 (Smith 1996, 303).

10 The puppet phenomenon originated in the work that the Bread and Puppet Theater did during the anti–Vietnam War movement.

11 Interview given to National Public Radio and reported on "Georgia Gazette," November 12, 1999.

12 Passage was a hard-liner who had extensive experience repressing national liberation struggles. At the beginning of his career, he worked with the U.S. military in Vietnam as a "pacification program analyst," and during the height of the civil war in El Salvador, he was Deputy Chief of Mission / Charge d'Affaires at the U.S. embassy in El Salvador. Passage also supported the Reagan administration's policy of "constructive engagement" with the apartheid regime in South Africa. His last administrative position was at the State Department, where he served as Director of Andean Affairs before his retirement in 1998.

13 Le Moyne was a colonel during the 1991 Gulf War when he commanded the Twenty-Fourth Division's First Brigade, and through the support of Twenty-Fourth Division commander General Barry McCaffrey, Le Moyne was promoted to general after the war. Subsequently, however, both Le Moyne and McCaffrey came under fire for their behavior in the war. The most serious allegation concerned the shooting of Iraqi prisoners by soldiers in the First Brigade, an event that Le Moyne is alleged to have covered up. See Hersh (2000).

14 USARSA was used by the military in lieu of the shorter SOA as a way to distance itself from the movement, which preferred SOA.

15 Goodman believed that movement leaders Roy Bourgois and Carol Richardson willfully manipulated college students, who, he felt, did not appreciate the complexities of the debate or fully understand the issues.

16 Final Report on the VIIth Meeting of the Board of Visitors, 1–3 February 2000. Ms.

17 Recent social-movement scholarship stresses that positing a theoretical opposition between "strategy" and "identity" inappropriately separates two aspects of all social movements that are mutually constitutive (e.g., Smith 1996).

18 Cited by Dominick (1999, 2).

19 Unable to completely distance himself from his military background, Blair saluted the crowd after addressing a group of protesters prior to the 2000 vigil.

20 Smith (1996, 211–364) describes how the Central American solidarity movement faced similar tensions over how broadly or narrowly to focus the struggle.

Conclusion

1 On Guatemala, see Godoy (2002) for a discussion of vigilantism as an outgrowth of the social trauma induced by state terror. For Honduras, the Comisionado Nacional de los Derechos Humanos (CNDH) has published various reports on rising crime and the associated extrajudicial executions of juvenile criminals by the police. See, for example, "Asesinatos de menores se duplican cada dos año," Boletin No. 1519, 25 June 2002, and "Armas de violencia en honduras," Boletin No. 1542, 30 July 2001. For Bolivia, Daniel Goldstein links the consolidation of neoliberalism to a rise in vigilante violence in a poor neighborhood of Cochabamba (Goldstein 2003).

2 Haitian paramilitary leader Emmanuel "Toto" Constance lives in the borough of Queens in New York City. Constance was also an informer for the Central Intelligence Agency. Along with other former Haitian military leaders who live in the United States, he has received protection from the efforts of groups who seek to hold him accountable for crimes committed in Haiti and deport him from the United States.

3 See Rosen (2002, 29).

4 See Ignatieff (2003).

5 See Ferguson (2003).

6 Telephone interview with Army Colonel Ken LePlante (ret.), 6 October 2002.

7 On Sunday evening, 22 September 2002, for example, the television news program *60 Minutes* engaged the question "Is torture justifiable?" and featured Harvard law professor Alan Dershowitz and a retired French general who fought in Algeria and used torture regularly against detainees. Both argued that under some circumstances torture is appropriate, and Dershowitz advocated the idea of "torture warrants" issued by judges. International law, however, prohibits torture under any circumstances.

8 See, for example, Van Natta (2003).

9 See Priest and Gellman (2002).

10 See Gumbel (2003).

11 Telephone interview with Colonel Richard Downie, 6 October 2002.

12 The Patriot Act of 2001, for example, makes it much easier for the FBI to spy on Americans by monitoring their telephone conversations and leisure activities. See EPIC (2002) for an analysis of the increasing surveillance, weakening data protection regimes, and heightened profiling of individuals since September 11, 2001, in the United States and other countries.

REFERENCES CITED

✦

Interviews by the Author

This is a partial list of interviewees. In some cases, interviewees asked not to be identified. In other cases, the identities of interviewees have been concealed in order to protect their safety. There are also a large number of people with whom I spoke repeatedly at the School of the Americas and in Bolivia, Colombia, Honduras, Columbus, Georgia, and Washington, D.C., but never formally interviewed. I have not included their names, as our exchanges took place during the activities in which I participated over the course of the research period. Persons are listed with the date and place that the interviews took place. All interviews were tape recorded unless noted otherwise.

SOA Officials

Bennett, Colonel Tom. 11 February 2000. School of the Americas. Not tape recorded.

DePalo, Colonel William (ret.). Commandant, January 1989–May 1991. 29 March 2001. Telephone interview.

Feliciano, Colonel José (ret.). Commandant, March 1991–March 1993. 9 November 1999. Columbus, Georgia.

Himes, Colonel Cecil (ret.). Commandant, July 1958–July 1964. 10 November 1999. Birmingham, Alabama.

Leur, Joseph. 28 September 1999. School of the Americas. Not tape recorded.

Loomis, Major Coredon, Jr. (ret.). Former instructor, public affairs officer, head of host family program. 11 November 1999. Columbus, Georgia.

Ruff, Lieutenant Colonel George. Director, SOA Department of Joint and Combined Operations. 11 February 2000. School of the Americas.

Santamaria, Walter. Translator. 9 February 2000. School of the Americas.

Sierra, Colonel Michael (ret.). Commandant, June 1984–October 1985. 6 April 2001. Fairfax, Virginia.

Trumble, Colonel Roy. Commandant, January 1995–July 1998. 8 November 1999. School of the Americas.

Weidner, Colonel Glen. Commandant, July 1998–January 2001. 29 September 1999 and 19 June 2000. School of the Americas.

Graduates of the Command and General Staff Officers Course

BOLIVIA

Calderón Terán, Colonel Fernando. 28 October 2001. Cochabamba, Bolivia.

Castellanos, General Ciro. 12 June 2001. La Paz, Bolivia.

Chulup Liendo, Colonel Guillermo. 19 June 2001. Cochabamba, Bolivia.

Delfin Mesa, Colonel Carlos. 19 June 2001. Cochabamba, Bolivia.

de la Fuente, Lieutenant Colonel Ramiro. 11 June 2001. La Paz, Bolivia.

García Rubin de Célis, General Victor. 12 June 2001. La Paz, Bolivia.

Pacello Aguirre, Lieutenant Colonel Oscar. 11 June 2001. La Paz, Bolivia.

Rojas, Lieutenant Colonel César. 13 June 2001. La Paz, Bolivia.

Sandoval Espinoza, Lieutenant Colonel Gustavo. 18 June 2001. Cochabamba, Bolivia.

Viduarre Noriega, Colonel Miguel. 19 June 2001. Cochabamba, Bolivia.

COLOMBIA

Arévalo García, Colonel Joaquín. 13 August 2001. Bogotá, Colombia.

Arteaga Arteaga, Major General Alfonso. 16 August 2001. Bogotá, Colombia.

Cáseres Carvajal, Rear Admiral Julio César. 14 August 2001. Bogotá, Colombia.

Delgado Caldas, Brigadier General Sigifredo (ret.). 15 August 2001. Bogotá, Colombia.

Flores Jiménez, Colonel Florentino. 15 August 2001. Bogotá, Colombia.

González Herrera, Major General Alberto. 15 August 2001. Bogotá, Colombia.

Ospina Galvis, Lieutenant Colonel Gustavo. 13 August 2001. Bogotá, Colombia.

Peña Porras, Lieutenant Colonel Héctor. 14 August 2001. Bogotá, Colombia.

Samaca Rodríguez, Colonel Héctor Julio. 15 August 2001. Bogotá, Colombia.

Sánchez Vargas, Major General Euclides. 13 August 2001. Bogotá, Colombia.

Urbina Sánchez, Brigadier General Luis Bernardo (ret.). 16 August 2001. Bogotá, Colombia.

HONDURAS

Arias Rodríguez, Colonel Juan (ret.). 31 May 2001. Tegucigalpa, Honduras.

Banegas Medina, General José de Jesus (ret.) May 30, 2001 Tegucigalpa, Honduras.

Bonilla Blanco, Colonel Rodolfo. 31 May 2001. Tegucigalpa, Honduras.

Díaz Zelaya, Lieutenant Colonel Fredy. 30 May 2001. Tegucigalpa, Honduras.

Estrada, Colonel Raúl. 1 June 2001. Tegucigalpa, Honduras.

Garcia Turcios, Colonel Abraham. 1 June 2001. Tegucigalpa, Honduras.

Maldonaldo, Colonel Luis A. 31 May 2001. Tegucigalpa, Honduras.

Maradiaga, Captain René. 1 June 2001. Tegucigalpa, Honduras.

Pineda Sagastume, Colonel René. 30 May 2001. Tegucigalpa, Honduras.

Romero, General Álvaro (ret.). 5 June 2001. Tegucigalpa, Honduras.

Rosales Abella, Brigadier General Marco Antonio (ret.). 5 June 2001. Tegucigalpa, Honduras.

Valencia García, Colonel Rolando (ret.). 4 June 2001. Tegucigalpa, Honduras.

Zelaya Reyes, Colonel Omar (ret.). 29 May 2001. Tegucigalpa, Honduras.

UNITED STATES

Rodríguez, Lieutenant Colonel Rand. U.S. Military Group, La Paz, Bolivia. 10 June 2001. La Paz, Bolivia.

Anti-SOA Activists

Bourgois, Father Roy. 20 September 1999. Repeated follow-up interviews, 2000–2002. Columbus, Georgia. Not tape recorded.

Liteky, Charlie. 20 September 1999. Columbus, Georgia.

Little, Ken. 19 June 2002. Telephone interview. Not tape recorded.

Lowe, Kate. 13 June 2002. Telephone interview. Not tape recorded.

Panetta, Linda. 29 May 2002. Philadelphia, Pennsylvania. Not tape recorded.

Reinbold, Leone. 20 June 2002. Telephone interview. Not tape recorded.

Richardson, Carol. 15 April 1998. Washington, D.C.

Taylor, Gail. 24 June 2001. Washington, D.C. Not tape recorded.

Triggs, Bruce. 19 June 2002. Telephone interview. Not tape recorded.

Wenders, Jeff. 24 June 2001. Washington, D.C. Not tape recorded.

Miscellaneous

Collins, Ron and Carol. SOA host family. 12 November 1999. Columbus, Georgia.

Cruz, Andrea. Director, Southeast Georgia Community Project. 17 November 2000. Lyons, Georgia.

Díaz Versón, Salvador. SOA support group. 16 February 2000. Columbus, Georgia.

Durán, Margarita. Leader, Bolivian coca growers. 17 June 2001. Chimoré, Bolivia.

Huanca, Casimiro. Leader, Coca Growers' Federation of Chimoré. 16 June 2001. Chimoré, Bolivia.

Gorman, General Paul (ret.). Commander, SOUTHCOM, 1983–85. 18 October 2000. Arlington, Virginia.

Laime, René. Director, Radio Soberanía. 17 June 2001. Chipiriri, Bolivia.

Morales, Evo. National Leader, Bolivian coca growers and MAS political party. 16 June 2001. Villa Tunari, Bolivia.

Rutledge, Olivia. Principal, Spencer High School. 17 November 2000. Columbus, Georgia.

Woerner, General Fred (ret.). Commander, SOUTHCOM, 1987–89. 11 November 2000. Alexandria, Virginia.

Published Materials

Agence France Presse. 2002. "U.S. Loses Battle against U.N. Anti-Torture Treaty," 8 November.

Alvarenga, Patricia. 1996. Cultura y ética de la violencia: El Salvador, 1880–1932. San José, Costa Rica: EDUCA.

Amnesty International, USA. 2002. Unmatched Power, Unmet Principles: The Human Rights Dimensions of U.S. Training of Foreign Military and Police Forces. Washington, D.C.: Amnesty International.

ANCD (Argentine National Commission of the Disappeared). 1986. Nunca Más. New York: Farrar, Straus, and Giroux.

Andersen, Martin Edwin. 1993. Dossier Secreto: Argentina's Desaparecidos and the Myth of the "Dirty War." Boulder: Westview.

Arendt, Hannah. 1994. Eichmann in Jerusalem: A Report on the Banality of Evil. New York: Penguin.

Bauman, Zygmunt. 1989. Modernity and the Holocaust. Ithaca: Cornell University Press.

Binford, Leigh. 1996. The El Mozote Massacre: Anthropology and Human Rights. Tucson: University of Arizona Press.

Browning, Christopher. 1992. Ordinary Men: Reserve Police Battalion 101 and the Final Solution in Poland. New York: Harper Perennial.

Burdick, John. 1998. Blessed Anastácia: Women, Race, and Popular Christianity in Brazil. New York: Routledge.

Bureau for International and Law Enforcement Affairs. 2001. International Narcotics Control Strategy Report. U.S. State Department, Washington, D.C.

Caldeira, Teresa P. R., and James Holston. 1999. "Democracy and Violence in Brazil." Comparative Studies in Society and History 41(4): 691–729.

Calhoun, Craig. 1997. Nationalism. Minneapolis: University of Minnesota Press.

Carbonella, August. 2002. "Memories of Imperialism: The Figure of the 'Vietnam' Veteran in the Age of Globalization." Ms.

Carmack, Robert M., ed. 1988. Harvest of Violence: The Mayan Indians and the Guatemalan Crisis. Norman: University of Oklahoma Press.

Castañeda, Jorge G. 1993. Utopia Unarmed: The Latin American Left after the Cold War. New York: Knopf.

Castellanos, Julieta. 2000. "Seguridad ciudadana, sociedad civil y respuesta institucional." In *Gobernabilidad democrática y seguridad ciudadana en centroamérica: El caso de Honduras*, ed. Leticia Salomón. Tegucigalpa: Centro de Documentación de Honduras. 87–117.

CDI (Center for Defense Information). 2002. Military Almanac 2001–2002. Washington, D.C.

Chevigny, Paul. 1995. *Edge of the Knife: Police Violence in the Americas*. New York: New Press.

Chomsky, Noam. 2000. *Rogue States: The Rule of Force in World Affairs*. Boston: South End Press.

Cohn, Gary, and Ginger Thompson. 1995. "Unearthed: Fatal Secrets." Special Report. Battalion 3-16. *Baltimore Sun*, 11–18 June.

Comisión de la Verdad. 1993. De la locura a la esperanza: La guerra de doce años en El Salvador. Informe de la Comisión de la Verdad. *Estudios Centroamericanos* 48, no. 533: 153–356.

CNDH (Comisionado Nacional de los Derechos Humanos). 2002. "Asesinatos de menores se duplican cada dos años." Boletin No. 1519, 25 June.

———. 2001. "Armas de violencia en honduras." Boletin No. 1542, 30 July.

CNPDH (Comisionado Nacional de Proteccíon de los Derechos Humanos). 1994. Los hechos hablan por sí mismos. Tegucigalpa: Editorial Guaymuras.

Conroy, John. 2000. *Unspeakable Acts, Ordinary People: The Dynamics of Torture*. New York: Knopf.

Cotler, Julio. 1983. "Democracy and National Integration in Peru." In *The Peruvian Experiment Reconsidered*, ed. Cynthia McClintock and Abraham H. Lowenthal. Princeton: Princeton University Press. 3–38.

Cunningham, Hilary. 1995. *God and Cesar at the Rio Grande*. Minneapolis: University of Minnesota Press.

Demarest, Geoffrey B. 1994. "Redefining the School of the Americas." *Military Review* 74(10): 43–51.

Dinges, John. 2004. *The Condor Years: How Pinochet and His Allies Brought Terrorism to Three Continents*. New York: New Press.

Dinges, John, and Saul Landau. 1980. *Assassination on Embassy Row*. New York: Pantheon.

Dominick, Brian. 1999. "Georgia on My Mind: Hard Thoughts on Closing the SOA. Ms.

Dorfman, Ariel. 2002. "Letter to America." *The Nation*, 30 September.

Dunkerley, James. 1984. *Rebellion in the Veins: Political Struggle in Bolivia, 1952–1982*. London: Verso.

Edelman, Marc. 1999. *Peasants against Globalization: Rural Social Movements in Costa Rica*. Palo Alto: Stanford University Press.

———. 2001. "Social Movements: Changing Paradigms and Forms of Politics." *Annual Review of Anthropology* 30: 285–317.

Enloe, Cynthia. 2000. *Maneuvers: The International Policies of Militarizing Women's Lives*. Berkeley: University of California Press.

EPIC (Electronic Privacy Information Center). 2002. Privacy and Human Rights: An International Survey of Privacy Laws and Developments. Washington, D.C.

Falk, Richard. 2001. "Defining a Just War." *The Nation*, 29 October.

Farthing, Linda. 1997. "Social Impacts Associated with Antidrug Law 1008." In *Coca, Cocaine, and the Bolivian Reality*, ed. Madeline Barbara Léons and Harry Sanabria. Albany: State University of New York Press. 253–70.

Feitlowitz, Marguerite. 1998. *A Lexicon of Terror: Argentina and the Legacies of Torture*. New York: Oxford University Press.

Ferguson, Niall. 2003. "No Way to Run an Empire," *New York Times Magazine*, 27 April.

Galeano, Eduardo. 2000. *Upside Down: A Primer for the Looking Glass World*. New York: Metropolitan Books.

Gill, Lesley. 1997. "Creating Citizens, Making Men: The Military and Masculinity in Bolivia." *Cultural Anthropology* 12(4): 527–50.

———. 2000. *Teetering on the Rim: Global Restructuring, Daily Life, and the Armed Retreat of the Bolivian State*. New York: Columbia University Press.

Go, Julian. 2000. "Chains of Empire, Projects of State: Political Education and U.S. Colonial Rule in Puerto Rico and the Philippines." *Comparative Studies in Society and History* 42(2): 333–62.

Godoy, Angelina Snodgrass. 2002. "Lynchings and the Democratization of Terror in Postwar Guatemala: Implications for Human Rights." *Human Rights Quarterly* 24(3): 640–61.

Goldstein, Daniel. 2002. "In Our Own Hands: Lynching, Justice, and the Law in Bolivia." *American Ethnologist* 30(1): 22–43.

Goñi, Uki. 2002. *The Real Odessa: Smuggling Nazis to Perón's Argentina*. New York: Granta Books.

Gootenberg, Paul, ed. 1999. *Cocaine: Global Histories*. New York: Routledge.

Gould, Jeffrey. 1990. *To Lead as Equals: Rural Protest and Political Consciousness in Chinandega, Nicaragua, 1912–1979*. Chapel Hill: University of North Carolina Press.

Gould, Stephen Jay. 1981. *The Mismeasure of Man*. New York: Norton.

Gourevitch, Philip. 1998. *We Wish to Inform You That Tomorrow We Will Be Killed with Our Families*. New York: Farrar, Straus and Giroux.

Grandin, Greg. 2000. *The Blood of Guatemala: A History of Race and Nation*. Durham: Duke University Press.

Green, Linda. 1999. *Fear as a Way of Life: Mayan Widows in Rural Guatemala*. New York: Columbia University Press.

Guillermoprieto, Alma. 2000. "Our New War in Colombia." *New York Review of Books*. April.

Gumbel, Andrew. 2003. "America Admits Suspects Died in Interrogations." *The Independent*, 3 March.

Gusterson, Hugh. 1998. *Nuclear Rites: A Weapons Laboratory at the End of the Cold War*. Berkeley: University of California Press.

Hardt, Michael, and Antonio Negri. 2000. *Empire*. Cambridge, Mass.: Harvard University Press.

Hayner, Priscilla B. 2001. *Unspeakable Truths: Confronting State Terror and Atrocity*. New York: Routledge.

Hersh, Seymour M. 2000. "Overwhelming Force." *The New Yorker*, 22 May. 49–81.

Hirschman, Albert O. 1988. "The Principle of Conservation and Mutation of Social Energy." In *Direct to the Poor: Grassroots Development in Latin America*, ed. Sheldon Annis and Peter Hakim. Boulder: Lynne Reinner.

Hobsbawm, Eric, with Antonio Polito. 2000. *On the Edge of the New Century*. New York: New Press.

Holden, Robert H. 1993. "The Real Diplomacy of Violence: United States Military Power in Central America, 1950–1990." *International History Review* 15(2): 221–440.

———. 1996. "Constructing the Limits of State Violence in Central America: Towards a New Research Agenda." *Latin American Research Review* 28(2): 435–60.

Hollander, Nancy Caro. 1997. *Love in a Time of Hate: Liberation Psychology in Latin America*. New Brunswick: Rutgers University Press.

Huggins, Martha K. 1998. *Political Policing: The United States and Latin America*. Durham: Duke University Press.

Huggins, Martha, Mika Haritos-Fatouros, and Philip Zimbardo. 2002. *Violence Workers: Police Torturers and Murders Reconstruct Brazilian Atrocities*. Berkeley: University of California Press.

Huntington, Samuel. 1981. *The Soldier and the State*. Cambridge, Mass.: Belknap Press.

HRW (Human Rights Watch / Americas). 1994. The Facts Speak for Themselves: Preliminary Report on Disappearances of the National Commissioner for the Protection of Human Rights in Honduras. New York: Human Rights Watch and Center for Justice and International Law.

———. 1996. Colombia's Killer Networks: The Military-Paramilitary Partnership and the United States. New York: Human Rights Watch.

———. 2001. The "Sixth Division:" Military-Paramilitary Ties and U.S. Policy in Colombia. New York: Human Rights Watch.

———. 2001. Human Rights Watch World Report. www.hrw.org / wr2kl / americas / colombia.html. Accessed 28 October 2001.

Ignatieff, Michael. 2003. "The Burden." *New York Times Magazine*, 5 January.

Isacson, Adam and Joy Olson. 2001. *Just the Facts, 2000–2001: A Civilian's Guide to U.S. Defense and Security Assistance to Latin America and the Caribbean*. Washington, D.C.: Latin America Working Group and Center for International Policy.

Jacobson, Matthew Frye. 2000. *Barbarian Virtues: The United States Encounters Foreign Peoples at Home and Abroad, 1876–1917.* New York: Hill and Wang.

Joseph, Gilbert. 1998. "Close Encounters: Toward a New Cultural History of U.S.-Latin American Relations." In *Close Encounters of Empire: Writing the Cultural History of U.S.-Latin American Relations,* ed. Gilbert Joseph, Catherine C. LeGrand, and Ricardo D. Salvatore. Durham: Duke University Press. 3–46.

Joseph, Gilbert, Catherine C. LeGrand, and Ricardo D. Salvatore, eds. 1998. *Close Encounters of Empire: Writing the Cultural History of U.S.-Latin American Relations.* Durham: Duke University Press.

Kaplan, Amy, and Donald E. Pease, eds. 1996. *Cultures of United States Imperialism.* Durham: Duke University Press.

Klare, Michael T., and Peter Kornbluh. 1988. *Low Intensity Warfare: Counterinsurgency, Proinsurgency, and Antiterrorism in the Eighties.* New York: Pantheon.

LAB-iepala. 1982. *Narcotráfico y política: Militarismo y mafia en Bolivia.* Madrid: iepala.

LAWG (Latin American Working Group). 1999. *Just the Facts: A Citizen's Guide to U.S. Defense and Security Assistance to Latin America and the Caribbean.* Washington, D.C.

Ledebur, Kathryn. 2002. "Coca and Conflict in the Chapare." *Drug War Monitor.* July.

Leech, Garry M. 2002. *Killing Peace: Colombia's Conflict and the Failure of U.S. Intervention.* New York: Information Network of the Americas.

Left Business Observer. 2002. "Reclaiming Empire." No. 102. 30 September.

Léons, Madeline Barbara and Harry Sanabria, eds. 1997. *Coca, Cocaine, and the Bolivian Reality.* Albany: State University of New York Press.

Levenson-Estrada, Deborah. 1994. *Trade Unionists against Terror: Guatemala City, 1954–1985.* Chapel Hill: University of North Carolina Press.

Lindsay-Poland, John. 2003. *Emperors in the Jungle: The Hidden History of the U.S. in Panama.* Durham: Duke University Press.

Linebaugh, Peter. 1992. *The London Hanged: Crime and Civil Society in the Eighteenth Century.* London: Cambridge University Press.

Lumpee, Lora. 2002. U.S. Foreign Military Training: Global Reach, Global Power and Oversight Issues. Special Report. *Foreign Policy in Focus.* May.

Lutz, Catherine. 2001. *Homefront: A Military City and the American Twentieth Century.* Boston: Beacon.

Madden, Teniente Coronel Patrick M. 2000. "La División de Officiales Internacionales." *Military Review* (Hispanic Version) (May–June): 19–23.

Malkin, Victoria. 2001. "Narcotrafficking, Migration, and Modernity in Rural Mexico." *Latin American Perspectives* 119(4): 101–28.

Manz, Beatriz. 1989. *Refugees of a Hidden War: The Aftermath of the Counterinsurgency War in Guatemala.* Albany: State University of New York Press.

McAdam, Doug. 1994. "Culture and Social Movements." In *New Social Move-*

ments: From Ideology to Identity, ed. Enrique Laraña, Hank Johnston, and Joseph R. Gusfeld. Philadelphia: Temple University Press. 36–57.

McClintock, Michael. 1991. "American Doctrine and Counterinsurgent State Terror." In *Western State Terrorism*, ed. Alexander George. New York: Routledge. 121–53.

———. 1992. *Instruments of Statecraft: U.S. Guerrilla Warfare, Counterinsurgency, and Counterterrorism, 1940–1990*. New York: Pantheon.

McSherry, Patrice. 1999. "Operation Condor: Clandestine Interamerican System." *Social Justice* 26(4): 144–74.

Meertens, Donny. 2001. "Facing Destruction, Rebuilding Life: Gender and the Internally Displaced in Colombia." *Latin American Perspectives* 28(1): 132–48.

Melguizo, Ramiro Ceballos. 2001. "The Evolution of the Armed Conflict in Medellín: An Analysis of the Major Actors." *Latin American Perspectives* 28(1): 110–31.

Menchú, Rigoberta. 1984. *I, Rigoberta Menchú*. London: Verso.

Milgram, Stanley. 1974. *Obedience to Authority: An Experimental View*. London: Tavistock.

Molano, Alfredo. 1992. "Violence and Land Colonization." In *Violence in Colombia: The Contemporary Crisis in Historical Perspective*, ed. Charles Berquist, Ricardo Peñaranda, and Gonzalo Sánchez. Wilmington, Del.: Scholarly Resources. 195–224.

Munczek Soler, Débora. 1996. *El impacto psicológico de la represión política en los hijos de los desparacidos y asesinados en Honduras*. Tegucigalpa: COFADEH and Comisionado de los Derechos Humanos.

National Security Archive. 2002. War in Colombia: Guerrillas, Drugs, and Human Rights in U.S.-Colombia Policy, 1988–2002. National Security Archive Electronic Briefing Book No. 69. Washington, D.C.

Nelson-Pallmeyer, Jack. 2001. *School of Assassins: Guns, Greed, and Globalization*. Maryknoll, N.Y.: Orbis Books.

Noriega, Manuel, and Peter Eisner. 1997. *Manuel Noriega: America's Prisoner*. New York: Random House.

O'Dougherty, Maureen. 2002. *Consumption Intensified: The Politics of Middle-Class Life in Brazil*. Durham: Duke University Press.

OMCT (Organización Mundial Contra la Tortura). 1992. *El terrorismo de estado en colombia*. Brussels: Ediciones NCO.

O'Shaughnessy, Hugh. 2000. *Pinochet: The Politics of Torture*. New York: New York University Press.

Owen, Roger, and Bob Sutcliffe. 1972. *Studies in the Theory of Imperialism*. London: Longman.

Panitch, Leo. 2002. "Violence as a Tool of Order and Change: The War on Terrorism and the Anti-Globalization Movement," *Monthly Review* 54(2): 12–32.

Panitch, Leo, and Sam Gindin. 2003. "Global Capitalism and American Empire," *The Socialist Register 2004*, 1–43.

Paul, Benjamin D., and William J. Demarest. 1988. "The Operation of a Death Squad in San Pedro La Laguna." In *Harvest of Violence*, ed. Robert M. Carmack. Norman: University of Oklahoma Press. 119–54.

Payne, Leigh. 2000. *Uncivil Movements: The Armed Right Wing and Democracy in Latin America*. Baltimore: Johns Hopkins University Press.

Pécaut, Daniel. 1999. "From the Banality of Violence to Real Terror: The Case of Columbia." In *Societies of Fear: The Legacy of Civil Violence and Terror in Latin America*, ed. Kees Koonings and Dirk Kruijt. London: Zed. 141–66.

Perera, Victor. 1993. *Unfinished Conquest: The Guatemalan Tragedy*. Berkeley: University of California Press.

Pérez, Louis A. 1995. *Cuba: Between Reform and Revolution*. New York: Oxford University Press.

Potter, George Ann, Ricardo Vargas, Melissa Drapper, Emilie Walker, and Emilio Ojopi. 2002 Rhetoric versus Reality: Alternative Development in the Andes (Bolivia, Colombia, and Peru). Reported submitted to the Drug Policy Alliance. Washington, D.C.

Priest, Dana, and Barton Gellman. 2002. "U.S. Decries Abuse but Defends Interrogations." *Washington Post*, 26 December.

Rabassa, Angel, and Peter Chalk. 2001. *The Colombian Labyrinth: The Synergy of Drugs and Insurgency and Its Implications for Regional Stability*. Santa Monica: Rand Corporation.

Radical History Review. 1993. "Imperialism: A Useful Category of Analysis?" Special Section. Vol 57: 1–84.

REMHI (Recovery of Historical Memory Project). 1999. *Guatemala: Never Again*. The Official Report of the Human Rights Office, Archdiocese of Guatemala. Abridged version. Maryknoll, N.Y.: Orbis Books.

Richani, Nazih. 2002. *Systems of Violence: The Political Economy of War and Peace in Colombia*. Albany: State University of New York Press.

Robben, Antonius C. G. 1995. "The Politics of Truth and Emotion among Victims and Perpetrators of Violence." In *Fieldwork under Fire: Contemporary Studies of Violence and Survival*, ed. Carolyn Nordstrom and Antonius Robben. Berkeley: University of California Press. 81–103.

Robins, Ron. 2003. *The Making of the Cold War Enemy: Culture and Politics in the Military Industrial Complex*. Princeton: Princeton University Press.

Roldán, Mary. 2002. *Blood and Fire: La Violencia in Antioquia, Colombia, 1946–1953*. Durham: Duke University Press.

Ronfeldt, David, John Arquilla, Graham E. Fuller, and Melissa Fuller. 1998. *The Zapatista Social Netwar in Mexico*. Santa Monica: Rand Arroyo Center.

Rosen, Stephen Peter. 2002. "The Future of War and the American Military." *Harvard Magazine* 104(5): 29.

Rosenberg, Tina. 1991. *Children of Cain: Violence and the Violent in Latin America*. New York: Penguin.

Sanabria, Harry. 1993. *The Coca Boom and Rural Social Change in Bolivia*. Ann Arbor: University of Michigan Press.

Schirmer, Jennifer. 1998. *The Guatemalan Military Project: A Violence Called Democracy*. Philadelphia: University of Pennsylvania Press.

Schulz, Donald E., and Deborah Sundloff Schulz. 1994. *The United States, Honduras, and the Crisis in Central America*. Boulder: Westview.

Scott, Peter Dale, and Jonathan Marshall. 1991. *Cocaine Politics: Drugs, Armies, and the CIA in Central America*. Berkeley: University of California Press.

Schoultz, Lars. 1981. *Human Rights and United States Policy toward Latin America*. Princeton: Princeton University Press.

———. 1987. *National Security and United States Policy toward Latin America*. Princeton: Princeton University Press.

Secretaría del Ejército. 1993. Operaciones Sicólogicas. Washington, D.C.

Sider, Gerald. 2000. "The Allies Salute You: Impunity, Slavery Work, and the Production of Differential Citizenship." Ms.

Sider, Gerald, and Gavin Smith. 1997. "Introduction." In *Between History and Histories: The Making of Silences and Commemorations*, ed. Gerald Sider and Gavin Smith. Toronto: University of Toronto Press. 3–30.

Silverstein, Ken. 2000. *Private Warriors*. London: Verso.

SIPRI (Stockholm International Peace Research Institute). 2002. *Yearbook 2002*. London: Oxford University Press.

Sluka, Jeffrey. 2000. "Introduction: State Terror and Anthropology." In *Death Squad: The Anthropology of State Terror*, ed. Jeffrey Sluka. Philadelphia: University of Pennsylvania Press. 1–45.

Smith, Christian. 1996. *Resisting Reagan: The U.S. Central America Peace Movement*. Chicago: University of Chicago Press.

Smith, Neil. 2003. *American Empire: Roosevelt's Geographer and the Prelude to Globalization*. Berkeley: University of California Press.

Spedding, Allison. 1994. *Wachu Wachu: Cultivo de coca e identidad en los yunkas de La Paz*. La Paz: HISBOL.

Stanley, William. 1996. *The Protection Racket State: Elite Politics, Military Extortion, and Civil War in El Salvador*. Philadelphia: Temple University Press.

Stoler, Ann Laura, and Frederick Cooper, eds. 1997. *Tensions of Empire: Colonial Cultures in a Bourgeois World*. Berkeley: University of California Press.

Tate, Winifred. 2001. "Into the Andean Quagmire: Bush II Keeps Up March to Militarization." NACLA *Report on the Americas* 25(3): 45–49.

Taussig, Michael. 1987. *Shamanism, Colonialism, and the Wild Man: A Study in Terror and Healing*. Chicago: University of Chicago Press.

Taylor, Michael. 2000. "A Matter of Honor." *San Francisco Chronicle*, 13 March.

Tilly, Charles. 1985. "War Making and State Making as Organized Crime." In

Bringing the State Back In, ed. Peter Evans, Dietrich Rueschemeyer, and Theda Skocpol. New York: Cambridge University Press, 169–188.

Torzano Roca, Carlos F. 1997. "Informal and Illicit Economies and the Role of Narcotrafficking." In *Coca, Cocaine, and the Bolivian Reality*, ed. Madeline Barbara Léons and Harry Sanabria. Albany: State University of New York Press. 195–210.

Triggs, Bruce. 2002. "Globalizing SOA Watch." In *The Global Activist's Handbook*, ed. Mike Prokosch and Laura Raymond. New York: Thunder Mouth's Press / The Nation Books. 48–53.

USARSA (United States Army School of the Americas). 1964. School of the Americas Catalog. Fort Gulick, Canal Zone: U.S. Army.

——. 2000. Course Catalog 2000 / 2001. Ft. Benning, Ga.: U.S. Army.

United States Department of State. 2001. International Narcotics Control Strategy Report. Bureau for International and Law Enforcement Affairs. Washington, D.C.

Uprimny, Rodrigo. 1995. "Narcotráfico, régimen político de prevención de la farmacodependencia en Colombia: 1986–1994." In *Drogas, poder y región en colombia*, ed. Ricardo Vargas. Vol 1. *Economía y política*. Bogotá: CINEP. 59–146.

Van Natta Jr., Don. 2003. "Questioning Terror Suspects in a Dark and Surreal World." *New York Times*, 9 March.

Vargas Meza, Ricardo. 1998. "A Military-Paramilitary Alliance Besieges Colombia." In NACLA, *Report on the Americas: Militarized Democracy in the Americas* 32(3): 25–27.

——. 1999. The Revolutionary Armed Forces of Colombia (FARC) and the Illicit Drug Trade. Policy Brief. Washington Office on Latin America.

Verbitsky, Horacio. 1996. *The Flight: Confessions of an Argentine Dirty Warrior*. New York: New Press.

Vilas, Carlos. 1996. "Prospects of Democratisation in a Post-Revolutionary Setting: Central America." *Journal of Latin American Studies* 28(2): 461–504.

Weil, Jim, and Connie Weil. 1993. *Verde es la esperanza: Colonización, comunidad, y coca en la amazonía*. Cochabamba: Los Amigos del Libro.

Williams, William Appleman. 1980. *Empire as a Way of Life: An Essay on the Causes of America's Predicament along with a Few Thoughts about an Alternative*. New York: Oxford University Press.

Wolf, Eric R. 1999. *Envisioning Power: Ideologies of Dominance and Crisis*. Berkeley: University of California Press.

Youngers, Colletta. 2003. "Drug Trafficking and the Role of the United States in the Andes." In *Politics in the Andes: Identity, Conflict, Reform*, ed. Jo-Marie Burt and Philip Mauceri. Pittsburgh: University of Pittsburgh Press.

Zinn, Howard. 2001. "A Just Cause, Not a Just War." *The Progressive*. www.progressive.org / 0901 / zinn1101.html. Accessed 24 May 2002.

INDEX

✦

Abrams, John, 222
Adams, James, 70
agriculture: agrarian reform, 27–28, 73, 172;
    farmworkers in Columbus, Georgia, 33;
    free-market policies and, 179; land con-
    flicts and, 172, 173, 181; multinational cor-
    porations and, 179, 181, 193–94; transpor-
    tation for, 188, 194; USAID (U.S. Agency of
    International Development), 46, 174, 222.
    See also coca cultivation; coca eradication
Allende Gossens, Salvador, 79, 80, 113
Alpírez, Julio, 6, 113
Alvarado, Napoleon, 250 n.28
Álvarez Martínez, Gustavo, 85–86, 89
American Family Life Assurance Company
    (AFLAC), 27, 28
American Foreign Intelligence Assistance
    Program, 49
American values: as civilized, 31–32;
    Columbus, Georgia, 20–21, 29, 30, 32, 33–
    34; consumerism as, 35, 36–37, 92, 101, 127,
    238; English language and, 30, 123–24;
    Euro-American SOA instructors and, 40–
    41; Host Family Program for CGS stu-
    dents, 37, 122–23; race and, 30, 32–33, 66–
    67, 69, 247 n.2; social class and, 30–31, 33–
    34; in Washington, D.C. trip, 150–54;
    wives of CGS officers on, 124–26. *See also*
    United States

Amnesty International, 151, 152–53, 157
Amos, Elena, 26–28
anti-SOA movement. *See* SOA Watch
Arce, Gómez, Luis, 104, 251 n.2
Arendt, Hannah, 58
Argentina: Gustavo Álvarez Martínez in, 86;
    Escuela Mecánica, 103; Andrés Francisco
    Valdéz and, 137; Hondurans' military
    training in, 83–84, 87; human rights viola-
    tions in, 6, 11–12, 86, 137; internal policing
    in, 61; LAGS trainees from, 65–69; Juan
    López Grijalva in, 86–87; Carlos Menem
    and, 11; Operation Condor and, 96; Plaza
    de Mayo demonstrations in, 15
Arias, Juan, 88–89
Army Foreign Intelligence Assistance Pro-
    gram (Project X), 49
AUC (Auto Defensas Unidas de Colombia),
    182, 183
AVANSCO, 11–12

Bámaca, Efraín, 113
Banzer, Hugo, 78, 96, 98, 135, 174, 204
Battalion 3-16, 6, 85, 86, 87, 155–56, 157, 250
    n.32
Battle of Seattle (November 1999), 229
Bauman, Zygmunt, 53–54
Bedoya Piaro, Harold, 250 n.28
Benitez, E. M., 69

Lesley Gill is an Associate Professor of Anthropology
at American University. She is the author of *Teetering on
the Rim: Global Restructuring, Daily Life, and the Armed
Retreat of the Bolivian State* (2000), of *Precarious
Dependencies: Gender, Class, and Domestic Service in Bolivia*
(1994), and of *Peasants, Entrepreneurs, and Social Change:
Frontier Development in Lowland Bolivia* (1987).

Library of Congress Cataloging-in-Publication Data
Gill, Lesley.
The School of the Americas : military training and
political violence in the Americas / Lesley Gill.
p. cm. — (American encounters / global interactions)
Includes bibliographical references and index.
ISBN 0-8223-3382-1 (cloth : alk. paper)
ISBN 0-8223-3392-9 (pbk. : alk. paper)
1. U.S. Army School of the Americas.   2. Imperialism.
3. United States—Military relations—Latin America.
4. Latin America—Military relations—United States.
I. Title.   II. Series.
U428.S83G54 2004
320.97'071'5—dc22   2003027511